THE SEVEN-STEP HOMESTEAD

the

SEVEN-STEP
HOMESTEAD

A Guide for Creating the
BACKYARD MICROFARM
of Your Dreams

Leah M. Webb

Storey Publishing

The mission of Storey Publishing is to serve our customers by
publishing practical information that encourages
personal independence in harmony with the environment.

EDITED BY Lisa H. Hiley and Carleen Madigan
ART DIRECTION BY Jessica Armstrong
BOOK DESIGN BY Stacy Wakefield Forte
TEXT PRODUCTION BY Liseann Karandisecky

Cover and interior photography by © Thomas C. Webb
Additional interior photography by © antonivano/stock
  .adobe.com, 215; © Asovereign83/Dreamstime.com,
  173; © Heather Kasvinksy, https://thisnoshtalgiclife
  .com, 152; © Igor/stock.adobe.com, 114; John Henry
  Nelson, 91; © Lost_in_the_Midwest/stock.adobe.com,
  219; © Megan Marie Weaver/stock.adobe.com, 213 b.;
  © SE Viera Photo/stock.adobe.com, 220 b.; © Tony
  Campbell/stock.adobe.com, 209

Homestead plans by © Scott Jessop
Additional illustrations by Ilona Sherratt © Storey Publishing,
  LLC, except Courtesy of Ark of Taste, 168

Text © 2023 by Leah M. Webb

Storey books are available at special discounts when
purchased in bulk for premiums and sales promotions as
well as for fund-raising or educational use. Special editions
or book excerpts can also be created to specification. For
details, please call 800-827-8673, or send an email to
sales@storey.com.

Storey Publishing
210 MASS MoCA Way
North Adams, MA 01247
storey.com

Storey Publishing, LLC is an imprint of Workman Publishing
Co., Inc., a subsidiary of Hachette Book Group, Inc., 1290
Avenue of the Americas, New York, NY 10104

ISBNs: 978-1-63586-411-3 (print); 978-1-63586-412-0
(ebook)

Printed in China by Toppan Leefung Printing Ltd.
10 9 8 7 6 5 4 3 2 1

Library of Congress Cataloging-in-Publication Data
  on file

To Mom for growing
food and flowers

To Dad for working with
tools and trailers

To my husband, TC,
for letting me take
over the lawn

CONTENTS

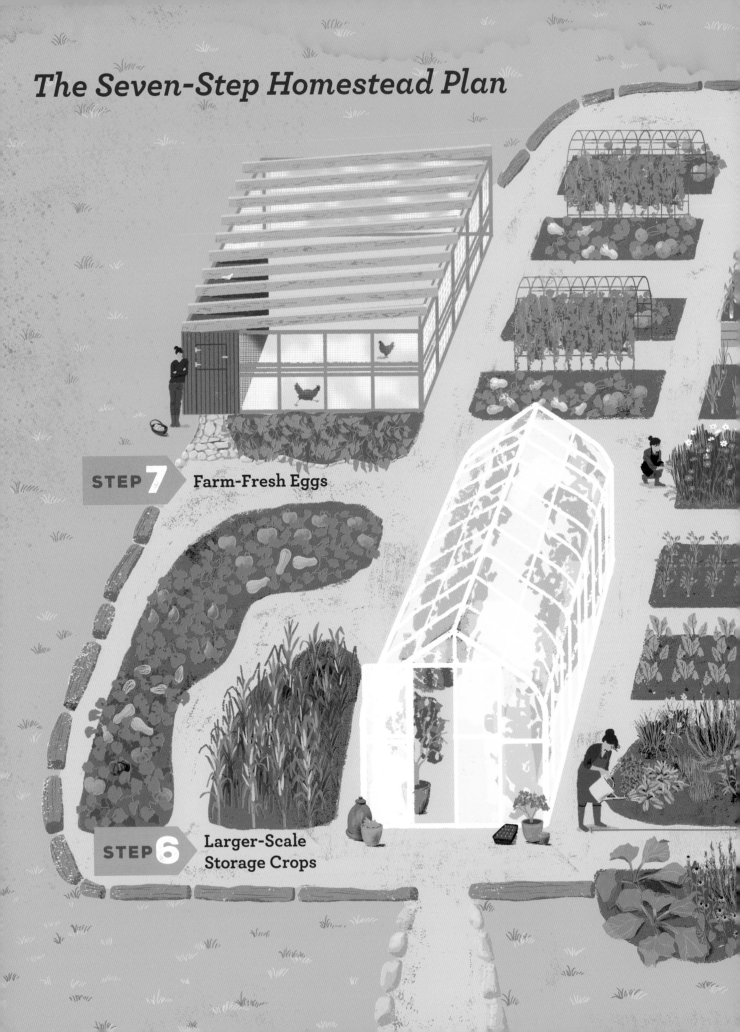

The Seven-Step Homestead Plan

STEP 7 Farm-Fresh Eggs

STEP 6 Larger-Scale Storage Crops

STEP 5 Four-Season Growing

STEP 3 Fruit Trees & Shrubs

STEP 2 Four Hundred Square Feet

STEP 4 Edible & Flowering Perennials

STEP 1 Starting Off Small: One or Two Beds

HOMESTEADING

it's about more than just food

I'M FORTUNATE TO HAVE always had gardens in my life. My mother was a gardener, but as a child I never saw the importance of her efforts nor understood the value in what she was doing. Now that I, too, am a mother, I've realized how lucky I was to grow up consuming the nourishment and knowledge that comes from homegrown foods. I assume it will be the same with my children in that it'll likely be decades until they know how privileged we were to have space to garden, money to buy compost, time to build the infrastructure, and the desire to watch it all grow.

My garden aspirations have evolved over the years. Cultivation started with a simple curiosity to grow a small portion of my own food. Heirloom tomatoes, fresh cucumbers, and abundant backyard greens had flavor like nothing I could purchase. They tasted so good, in fact, that I started to resent the lackluster quality of foods available at the grocery store. Motivated by culinary appeal, I began to expand my garden year after year. I had tasted the best, and I wanted more.

But growing food wasn't just about the taste. Superior nutrition was the driving force behind that superior flavor. I recognized that my homegrown foods provided a massive health benefit compared to conventionally grown foods that had often traveled thousands of miles before arriving at the market. And then there were the hours I spent outdoors and the movement involved in vegetable cultivation—the bending, squatting, leaning, carrying, digging, and other agilities required of a gardener. From a holistic health standpoint, organic gardening was a valuable tool to improve personal health.

The fuel and chemical inputs required by large-scale farming also informed my desire to grow my own food. Pesticide usage, operation of heavy farm equipment, plastic packaging, and cross-country transportation mean that the majority of food production poisons the earth that sustains us. By gardening, I could ever so slightly offset the environmental consequences of farming every time I harvested a meal from my backyard. I found hope in the fact that I could help store carbon by collecting dried leaves, twigs, and other organic waste products to build healthy soil. I wasn't just eliminating a portion of my carbon footprint; I was actually tipping the scales in the other direction!

> *My garden aspirations have evolved over the years. Cultivation started with a simple curiosity to grow a small portion of my own food.*

Teaching my children to enjoy healthy foods became another motivator. I remember reading in Barbara Kingsolver's *Animal, Vegetable, Miracle* of her husband's tale of a child who thought spaghetti was a root crop. At the time, this association sounded absurd, but my work in environmental education and as a volunteer in my children's schools proved this misconception and others like it to be all too common. I've met youngsters who wouldn't touch a turnip out of fear, let alone try tasting it.

Gardening creates a complete shift in attitude. Over the course of two years of managing a children's learning garden, I witnessed kids argue over the last Brussels sprout, casually shovel handfuls of lettuce into their mouths, and create sorrel "burritos" stuffed with strawberries and fennel greens. Gardening (and cooking, for that matter) provokes a child's natural curiosity and encourages healthful habits with little to no discussion about health. Teaching youth about "good" and "bad" foods requires a nuanced perspective (highly processed foods aren't so bad when that's all a family has to eat). Showing a child that vegetables can be fun first and delicious second is a more relatable message for a developing mind.

Growing your own food also provides a sense of security. I certainly wouldn't consider myself to be a prepper, but I do dabble in preparing for the unexpected. As history has shown us time and again, people turn to gardening during times of turmoil such as war, economic hardships, and, as we saw in 2020, pandemics. In my own situation, even with large gardens and many laying hens, I'm far from producing all my family's food. Fats, oils, meats, potato chips, ice cream, and other items on my grocery list are foods that would require even more work (and land) to make than I'm willing to invest. But should these items become unavailable someday, I'll be grateful that I've created at least some level of self-sufficiency.

I'm also tired of my dollars making the rich grow richer. My reliance on Big Ag only pours more money into the pockets of those who control the market. Growing my own food isn't cheap, but at least it's redirecting money away from a monopoly. Investing in small seed companies, local compost and soil, and goods from farmers' markets supports blue-collar workers whose income bracket is more similar to my own. My miniscule contribution is far more valued by small businesses than it is at the chain store where I buy my chips.

Needless to say, not everyone can grow food. Access to the benefits I've listed—personal and environmental health for kids and adults, self-sufficiency, and building local economy—is available only to those of us who possess the time, energy, money, and space to grow. In that sense, backyard gardening is a privilege and by no means represents a universal antidote when it comes to solving the health problems of a developed nation. But it does represent a small solution for those of us who are interested in taking a more active role in our food production.

As someone who relies on the foundational health benefits of gardening for both mental and physical well-being, I'm fully convinced of the tremendous opportunity for a sense of connection that we gain from cultivating the land, no matter how big or small of an effort. You may have picked up this book because you want to grow food, but my hope is that you find the joy, bewilderment, successes, and failures just as gratifying as that homegrown tomato. You're unlikely to find another hobby with so much to offer.

HOW TO BE A SUCCESSFUL GROWER (A.K.A. HOW TO USE THIS BOOK)

My goal for you is that gardening successes outweigh the failures, hopefully by a lot, every single year. For this to happen, you must trust me when I tell you to take it slow. Cultivating food seems as if it should come naturally. After all, humans have been eating and growing food for thousands of years. How hard could it possibly be?

Developing a green thumb (notice I said "developing," which doesn't imply that we are born with green or brown thumbs) takes practice. Developing productive beds takes time and resources. Despite your eagerness to start growing as much as you possibly can in whatever amount of space you've set aside, slow down and start small. Get to know your property, the angle of the sun throughout the seasons, the way your soil drains after heavy rains, what your average high and low temperature range is. Think about how you interact with your space and learn about local resources. When you start too big, you're spreading a finite amount of energy over a large amount of space. You're bound to miss important details that could have informed better growth. Every season, observe which varieties thrive and which fail, which pests are the most problematic, and which planting windows are most successful.

Most gardening books are organized by topic, with soil covered in one chapter, cover crops and mulches in another, planting in yet another. This book differs in that it organizes much of the same information in steps based on the gardener's experience and ability level, with the assumption that every year will see an expansion of growing space and an increase in the number and variety of crops. Step 1 teaches you how to build your beds and to grow the easiest plants. Step 2 expands on that information and introduces a few more techniques and a handful of more challenging plants. Each chapter thereafter includes just the information you need to successfully complete that particular step. This book is designed to teach you as you go and to help you expand your efforts with just the right amount of knowledge.

I've essentially broken my own homesteading efforts into seven distinct steps. If you follow the recommendations within, you'll wind up with about 1,800 square feet of diverse growing space, a small greenhouse, a small orchard, and a starter flock of chickens. My own garden is actually much larger than this because I've invested more energy in the facets of gardening I find most appealing.

I've also been doing this for a while. I would have been overwhelmed had I tried to build all my infrastructure in just a year or two. Taking 5 to 10 years is far more realistic. Furthermore, you're not just building infrastructure: You have to maintain it all and harvest and preserve your bounty. You might not realize how much work you're piling on until you're in the thick of it, or you may find that your initial goal was too lofty—a lesson better learned when your garden is small

I would have been overwhelmed had I tried to build all my infrastructure in just a year or two. Taking 5 to 10 years is far more realistic.

than after you've invested hundreds of dollars and hours in something that doesn't bring you joy.

The steps don't necessarily have to be followed in order, although you should absolutely start with Steps 1 and 2 if you're new to no-till gardening. You can skip around if you see a chapter that feels more exciting and come back to earlier chapters when you have interest. The order I've presented makes the most sense from a production, logistical, and learning perspective, but it may be that another order feels more appealing.

Whichever order you use to approach garden expansions, don't skimp on the initial time investment, make sure you're confident and comfortable in your efforts before moving on, and, whatever you do, never take on so much that you lose the excitement and joy of cultivating nourishment.

1

starting off small:

ONE OR TWO BEDS

I get it; you're eager to fill your table and pantry with fresh fruits and vegetables harvested from your own garden. How hard can it possibly be? You see no reason to start small! Yet no one ever picks up a musical instrument for the first time and expects to play a complex melody. Why would we assume that gardening is any different? We first learn foundational skills, work to master the easy stuff, then expand when we're ready.

You'll be surprised by all that can be produced in a small, well-tended garden.

the plan

- **Make just a couple of beds.** For your first garden, I recommend having one or two 4 × 8-foot beds. Concentrating your efforts in a small area increases your chances of success, which will inspire you to keep gardening. Over time, you'll learn to become more efficient, which helps you expand your garden without becoming overwhelmed.

- **Create no-dig beds using sheet mulching.** This simply means you build a bed from layers of organic material, for example, sticks, dried leaves, manure, mushroom compost, kitchen scraps, cardboard, paper, wood chips, wood shavings, or numerous other yard-waste products. It's a technique that results in a mounded bed that—importantly—doesn't disturb soil the way tilling does.

- **Choose the easiest plants to grow.** Small beds can accommodate only a certain number of plants, so it's best to select plants with the highest chance of success. The initial techniques for tending to these plants are simpler than the methods in later chapters, but they form the foundation for all stages of garden expansion.

Your most suitable site may require removing a few large limbs or even whole trees to improve sun exposure—tasks that are far easier to accomplish before establishing your garden. We removed four large trees before beginning bed construction.

BEFORE YOU DIG IN

As tempting as it is to get your hands in the soil as soon as possible, pause for a moment, think long term, and choose a site that will serve you for years to come. A successful garden requires adequate sun, proper drainage, and healthy soil. Considering all of these factors at once may feel overwhelming, and depending on your particular situation, it may seem that the perfect site doesn't exist. The key is to prioritize as many of the following criteria as you can while accepting your location's limitations. Besides, you can't know the advantages and limitations of a site until you've gardened there for a few years.

Choose the best site. Select an area that can be easily expanded in the future as your garden increases in size, productivity, and diversity. An ideal site has freely draining soil and sunlight for at least six, and ideally eight or more, hours a day. It should be within reach of a hose, visible from your house, and easily accessible. The leach fields of septic systems are unsuitable for vegetable garden sites but can be planted instead with shallow-rooted ornamentals (deeply rooted plants such as trees and shrubs can grow into and clog drain lines). In areas that contain buried utility pipes or power lines, do some research regarding pipe location and depth; you don't want to dig where pipes or lines are near the surface.

Observe your potential site(s) over the course of a few weeks or, even better, a few seasons if you're feeling patient. When it rains, where does the water go? How long does it take for the ground to dry out? Which direction does the sun move, and how much sun exposure do you get? (Keep in mind that the sun takes a slightly different path—resulting in shorter days—during winter months.) Are you able to see the garden from a window or when you immediately step outside your door? Gardens that are out of sight are often neglected!

Get to know your soil. Using a shovel, dig down 8 to 12 inches in the proposed garden site and see what lies below. Is the soil brown, gray, red, gravelly, compacted? By digging belowground, you have a chance to notice any major red flags: things such as erosion-control netting just under the surface, questionable fill (such as coal ash, gravel, or construction waste), or other surprises that will affect garden size and placement.

Think about sun requirements. How much sun you need depends on which crops you'll be growing. You can grow leafy greens with as little as five or six hours of sun per day, but fruiting crops such as tomatoes, peppers, cucumbers, and squash require eight or more hours of daily sunshine. Root vegetables such as carrots and beets lie somewhere between the greens and the fruits, but more sun is generally better. Crops that don't receive enough sun grow slowly and produce less overall.

Shady gardens are also slower to dry out. This may be a benefit in hot, sunny climates (even heat lovers like melons and peppers have their limits!), but it creates a breeding ground for fungal and bacterial infections in regions with heavy rainfall. Growers in milder climates or those with lower annual temperatures need to maximize their sun exposure to capture more heat.

KNOW YOUR USDA HARDINESS ZONE

The United States is divided into 13 different grow zones based on average annual minimum temperatures. Your zone tells you a lot about what you can or can't grow and when you should or shouldn't plant. This becomes particularly important when you start learning about perennials.

Regions that are classified as Zone 1, such as Alaska, experience temperatures as low as −60° to −50°F (−51° to −46°C). Zone 13 exists

TEST SOIL DRAINAGE

>>> The easiest way to test for drainage is to dig a hole about 12 inches deep and 12 inches wide. Fill the hole with water and time how long it takes for the water to soak into the soil. Perform this test immediately after a soaking rain to observe how water behaves when the surrounding soil is saturated. Alternatively, you can thoroughly water the hole and surrounding area for 5 minutes before performing the test. Let the water soak in for 20 minutes, then repeat the watering two or more times until the soil appears saturated.

Either way, the hole should drain within three hours. If water remains, you need to choose a different site or construct raised beds to keep plant roots from being sodden.

on the opposite end of the scale—these regions are closest to the equator where lows range from 60° to 70°F (16° to 21°C).

The letter "a" or "b" at the end of a zone number designates the upper or lower end of that zone. For example, I'm located in Zone 7a, which is slightly warmer than Zone 6b but colder than Zone 7b. There are several online resources that give hardiness zones; I like the app used by the National Gardening Association.

TILLING IS UNNECESSARY

Tilling soil is a highly debated topic, and understandably so, but I don't till—and I recommend that you don't either. Humans have been tilling soils since the invention of agriculture, so you might think that it makes sense to keep doing it. Yet in most cases, tilling simply isn't necessary for plants to grow well, and it comes with several downsides.

Roots are powerful. They generate enough pressure through their cells to grow through concrete and asphalt, so you know they can penetrate untilled soil. Although it's important to aerate soil and keep it from becoming compacted, we don't need to run a rototiller through garden beds every year to accomplish that. Earthworms and other soil biota will do this work for us by creating underground tunnels for air exchange and water infiltration.

Soil invertebrates like worms and other decomposers transport nutrients throughout the soil profile. Bacteria and fungi exude organic acids that encourage soil particles to bind together. The action of invertebrates combined with these sticky acids provides soil with a structure capable of holding water and nutrients. Tilling breaks apart the soil in a way that alters these processes and encourages nutrient loss.

Soil disturbances (tilling, double-digging, or other methods intended to break up clumps of soil) are essentially wounding the earth, and those nuisance weeds are nature's way of slapping on a Band-Aid. Tilling churns up weed seeds and deposits them on the surface where they can germinate. By opting not to till and by keeping soil covered with a mulch—living or

TO KEEP A GARDEN JOURNAL OR NOT

>>> Most garden advisors recommend record keeping as a key component in learning to garden. In 20 years, I've cumulatively filled one, maybe two pages with a few notes that have yet to inform any future decision. For me, record keeping steals some of the enjoyment of gardening. I'm certain that my efforts would improve if I were to make notes of my failures and successes, but it just isn't my style.

I can't necessarily recommend you follow my haphazard approach, however. I've grown varieties that I'd like to plant again but have no idea of its name or when it went in the ground. I've experimented with varieties that I'd prefer not to waste time on again, yet they mysteriously reappear.

Year-round gardening efforts (covered in Step 5) benefit most from good notetaking. Timing and variety selection can make or break your fall and winter garden. And because it takes a few tries to get year-round growing right, a handful of notes is a valuable guide when trying again the following season. Keeping a journal is not a requirement for success. But if record keeping appeals to you, start right from the beginning by recording varieties, planting dates, failures, successes, and any other important notes throughout the season.

not—you'll reduce the weed pressure. Adding a layer of compost to the soil surface (a practice known as topdressing) every year is another way to minimize weeds by smothering weed seeds at the soil surface.

ENRICH YOUR SOIL

Soil provides minerals and other important nutrients to plants, but these molecules need somewhere to be collected and stored. Enter carbon. Carbon is the basic building block of all life on Earth; every living and formerly living thing contains loads of it. When living things die and are decomposed by bacteria, fungi, and other organisms, the carbon molecules are broken into simpler molecules that act as storage units for nutrients. The process of decay unlocks these nutrients, making them available for plant uptake.

Compost, manure, leaves, and other forms of decaying organic matter are essentially heaps of carbon. When you add organic matter to the soil, you're adding the nutrition that's available in that decaying matter, but you're also adding a storage unit—carbon—that can harness even more nutrients. For example, nitrogen is a finicky molecule that exists in air and soil. It moves between the two quite readily but is more easily captured in soil when there's adequate carbon.

Organic matter also builds habitat for important organisms that further enhance soil nutrition. You're basically throwing a carbon party and inviting all the most important guests. Worms, fungi, bacteria, and a whole slew of lesser-known players in the soil ecosystem like to munch on carbon. Their excrement and other by-products of metabolism help hold soil particles together and act as an excellent source of plant nutrition.

TO TEST OR NOT TO TEST?

>>> Many gardeners recommend testing your soil, but I don't think it's necessary. Soil chemistry is complicated, and soil testing is often used by farmers to inform optimal growing conditions for just one or two crops, not the diverse range of crops in the home garden. Deficiencies flagged by lab tests are generally corrected in the same ways—by adding compost and organic matter, implementing no-till practices, planting cover crops, and applying varying mulches, which are all techniques covered in this book.

Measuring soil health requires a lot more than obtaining ideal nitrogen and phosphorus levels. The presence of nutrients in soil doesn't always mean those nutrients are available for plant uptake, so you also need to know how those nutrients are stored and whether they're available to the plants.

If you follow the instructions in this book and add compost or aged manure to your garden every year, you can improve your soil over time without the aid of soil test data. However, if you've done everything to properly manage the soil and you're still having failure after failure, then you may want to explore soil tests.

One instance in which I do recommend soil testing is when you suspect hazardous waste or other undesirable contaminants may be on the site where you plan to establish your garden. Some contaminants are visible—coal ash, for example, is unfortunately frequently used as fill dirt during new construction. If you can see contamination, seek advice from your local Cooperative Extension. But if you're planning to use a space that has no history of contamination, there's no pressing need to test the soil before beginning.

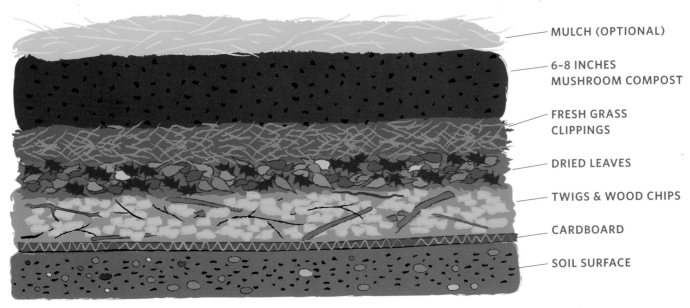

- MULCH (OPTIONAL)
- 6-8 INCHES MUSHROOM COMPOST
- FRESH GRASS CLIPPINGS
- DRIED LEAVES
- TWIGS & WOOD CHIPS
- CARDBOARD
- SOIL SURFACE

Sheet mulching is highly flexible and can be built with fewer layers than shown here. Start by smothering the grass with cardboard, then pile layers of organic matter on top.

BUILD YOUR BEDS

I recommend no-dig beds created by sheet mulching—layering organic matter directly on the soil. This simple and effective technique results in a mounded bed that improves drainage while keeping the beds close enough to the ground to retain enough moisture for proper plant growth.

When sheet mulching, you're essentially building rich, healthy growing medium on top of existing soil. I like to think of it as establishing a compost heap directly on the area that will become my garden. The first layer consists of thick cardboard to smother existing vegetation. The next comprises a variety of large carbon-rich waste products such as sticks, decaying wood, or untreated carpentry scraps. Then comes a thick layer of dried leaves, manure, mushroom compost, kitchen scraps, cardboard pieces, shredded paper, wood chips, and/or wood shavings. These smaller pieces of organic matter fall into the cracks between the larger pieces, reducing large air pockets and filling the bed.

As you assemble these layers, scatter some blood meal or other organic, nitrogen-rich fertilizer throughout to boost the bed's nutrition. Over time, this heap of carbon decays (the process is accelerated when additional nitrogen is added), inviting earthworms, fungi, bacteria, and other organisms into the beds. You end up with beautifully dark and rich soil right where you need it.

Before you put in plants or seeds, the final step is to smother the assembled layers with a thick blanket of compost, soil, or aged manure. This step is critical: It ensures that plants have a place for their roots to take hold. Although the bed's lower layers will decompose over time, they

Sheet-mulched beds can be built directly on the ground without a border as was done here.

A WORD OF CAUTION: HAY, STRAW, AND BROADLEAF HERBICIDES

>>> Hay and straw—and the manure or bedding from horses, cows, and other farm animals—are excellent additions to a garden (hay and straw should be well watered for a few weeks before placing in a garden so that any seeds have a chance to germinate and die). However, it's imperative to ensure that the hay or straw was not sprayed with a broadleaf herbicide such as picloram, clopyralid, or aminopyralid. Herbicides break down over time in the presence of sunlight, microbes, heat, and moisture, but some broadleaf herbicides can persist for as long as three to four years, even after being digested by animals.

Although these herbicides don't affect all plants, they make the soil unsuitable for growing carrots, peas, eggplants, tomatoes, peppers, beans, and other vegetables. I've known gardeners who made the mistake of using manure from animals that were fed treated hay, and the results were truly disappointing, especially since stunted vegetable growth can persist for years. If this happens, the only options are to remove as much of the soil as possible and replace it with new, uncontaminated organic matter, or to repeatedly plant cover crops on the beds for the next few years while you wait for the pesticides to degrade.

The manure that I get from my local beef farmers is a safe bet, as they can trace and verify the origins of their hay. Securing herbicide-free straw for use as mulch in the garden is slightly more challenging, but I'm able to source no-spray straw twice per year from a local feedstore. I buy 10 bales at a time and store them in the garage. Just one or two bales would likely fulfill the annual needs of a small garden.

offer minimal nutrition to plants for the first year or two. Adding a topcoat of a growing medium kick-starts the decomposition process in the bed's first season. This final topping needs to be nutritious if your garden is to flourish—I prefer to use mushroom compost. It's available in bulk from landscaping supply stores.

Invest the time and patience up front, and I guarantee that you'll set yourself up for better long-term gardening success.

ARE BED BORDERS NECESSARY?

Devotees of raised beds—beds built with borders—assert that they're the best way to garden. Raised beds do have a few bonuses: They look tidy, keep soil contained, prevent weeds from creeping in the sides of the beds, and drain well. Higher beds are easier to access for planting and harvesting, a consideration for gardeners who have difficulty bending over or kneeling. I use them in my front yard, which has compacted red clay soil. The frames are filled with rich organic matter that keeps plant roots away from the soggy soil below. I've also lined the bottom of each raised-bed frame with hardware cloth to prevent voles from burrowing under the bed and eating plant roots (see page 72).

Raised beds have some downsides, though. Because they drain so well, they're not ideal for hilltops or sunny, dry sites. And between the cost of the materials to build the bed frame and the mix to fill it, raised beds can get expensive. One option, if you decide you want borders, is to find free local materials instead of buying them; I use any straight tree with an 8- to 12-inch diameter that happens to fall on our property during a storm. Other materials to construct border frames include large rocks, cinder blocks, logs, wooden planks, or sheets of corrugated metal.

Numerous methods exist for building raised beds, and you can customize them for your needs. However you decide to build your beds—bordered or not, deep or shallow—the sheet-mulching method can be applied to all situations.

NO BROWN THUMBS

>>> I don't believe in brown thumbs; I believe that people give up prematurely. I have been gardening for 20 years, worked as a landscaper for 8 years, studied plant and soil science in college, and have been teaching gardening for the better part of 9 years. Do you know what I do every single year? I screw something up and then have to wait an entire year to try again. When you think about gardening with perspective, you realize it takes time to learn how to get it right.

Having some success early on will get you excited about gardening; starting with reasonable expectations and a small, manageable garden puts you on the right path. Focus on learning the basic skills well so that you can establish a solid knowledge base and develop your confidence.

Some people throw in the trowel because they feel like the garden is out of control. It helps to think of yourself as a facilitator rather than trying to manage everything that happens in the garden. Unexpected circumstances are guaranteed. Your job is to try to plan for the unexpected and accept that which is out of your control. I honestly can't imagine a better training ground for life.

Sheet Mulching

The steps may look official, but please realize that there's no wrong way to do this. My beds are often started with just a few ingredients—a layer of cardboard piled with wood chips, blood meal, and mushroom compost. Additional ingredients are a bonus. Walkways can be made with just the cardboard and a thick layer of wood chips.

BOTTOM LAYER

- Sheets of heavy cardboard

Apply two or three layers of cardboard. Overlap pieces by at least 8 inches, ideally 12, and cover any holes in the cardboard with additional layers to prevent sunlight from reaching the grass.

Don't skimp on the amount of cardboard; lighter layers won't kill the grass, meaning the grass will eventually grow through the organic matter you pile on top of the cardboard.

Alternatively, you can use just dried leaves to smother grass, but to be effective, the leaf layer must be 18 to 24 inches thick before you pile on the organic material in the next steps.

FILLER

- Twigs and trimmings from shrubbery
- Wood chips
- Shredded paper (nothing glossy)
- Partially decomposed kitchen waste
- Untreated packing paper and paper bags
- Corrugated cardboard scraps
- Compost or soil to fill in gaps

Pile the largest pieces of organic matter onto the cardboard first. Add 4 to 6 inches of smaller pieces of organic matter, shoveling them into the spaces and cracks of the first layer.

NITROGEN LAYER

- Blood meal (one of the most concentrated forms of organic nitrogen; easy to apply)
- Manure (especially from poultry)
- Grass clippings (fresh, chemical-free)

Adding nitrogen to the carbon-rich layers can accelerate the decomposition process and build richer soil quicker. Apply 2 to 3 inches of manure, or follow the manufacturer's instructions for applying granular fertilizers like blood meal.

PLANTING MEDIUM

- Mushroom compost
- Compost from citywide composting services
- Homemade compost from kitchen scraps and yard waste (must be fully composted for use as a top layer)
- Bagged or bulk organic compost or manure
- Aged manure

Finish the bed with 6 to 8 inches of mushroom compost, soil, or other organic growing medium.

Optional: Cover the bed with a thin layer of mulch such as straw, pine straw, grass clippings, dried leaves, or aged wood chips (not fresh wood chips; see page 71 for more about this). Mulch encourages moisture retention, which is important for plant growth and organic matter decomposition.

FINDING CARDBOARD

You'll be surprised at how easy it is to obtain loads of corrugated cardboard for use in the garden. Large, thick boxes such as those used to ship bikes and appliances are ideal, but any corrugated cardboard will work. You might have cardboard boxes on hand from online shopping deliveries, and neighbors and friends might be happy to give you theirs. Hardware stores, bike shops, restaurants, grocery stores, and many retail businesses generate a shocking amount of cardboard daily. Check with local stores to see whether they're willing to share. Avoid glossy cardboard, however, as the dyes can contain toxins.

SOURCING OTHER MATERIALS

The flexibility of sheet mulching is one of its finer attributes. There's no need to follow a prescriptive approach when constructing no-till beds because so many ingredients work perfectly well. Besides, sourcing organic matter in my region is likely to be a different process than in another part of the country. Use what's available and, better yet, look for material that's free! Make friends with landscapers and arborists, who might be looking to get rid of yard waste.

Ask carpenters and contractors if you can collect wood scraps, wood chips, and sawdust (nothing with paint, stains, or other chemical treatment). Arrange to pick up bags of leaves, grass trimmings, and other organic material from neighbors. Contact farms with chickens, horses, or cows—they might be delighted to have you haul away a few loads of manure.

If you approach bed construction with a bit of creativity and flexibility, you can drastically reduce the cost of construction while creating a nutrient-dense growing medium generated from products that may have otherwise ended up in a landfill.

There is one caveat: Avoid ingredients that contain dyes, herbicides, or other contaminants. As mentioned, I avoid glossy cardboard and do my best to remove labels or packing tape before placing cardboard in my beds. If collecting grass clippings from a neighbor, think twice if their grass seems just a bit too green or isn't at all "weedy." An overly manicured lawn may have been doused with broadleaf herbicides and chemical fertilizers. You don't want these synthetic compounds in your vegetable garden; at the very least, they can stunt plant growth; at worst, they can lead to human health problems.

CHOOSE YOUR PLANTS

For your first garden, I recommend using starts (plants that are grown in pots and planted as seedlings) for most crops and choosing a few plants to direct sow (planting seeds in the ground) for practice. Purchase the starts from local growers, farmers' markets, and small nurseries to ensure that you plant varieties suited to your region.

Avoid purchasing starts too early or before the beds are ready, as you'll be responsible for keeping them alive and well. If you do purchase starts early, be sure to harden them off before planting (see page 76). Once you've expanded your garden in future years, you may want to start your own plants from seed, both for the cost savings and for the wider variety of plants than what's available as starts.

Compared to using starts, direct sowing in the garden can be a bit challenging, for several reasons: First, some seeds have finicky germination requirements. Second, direct sowing requires greater attention to watering while the seeds germinate and establish themselves; both over- and under-watering are problematic. Third, pests may quickly devour seedlings when they emerge from the ground (plants can survive the same level of herbivory once they're larger and well established).

EASY PLANTS FOR BEGINNERS

I always recommend that new gardeners limit their crop selection to those that are easiest to grow. The following plant recommendations are ideal for someone who has little to no gardening experience. You'll notice that some of these plants can either be started from seed or from starts; they are all fairly easy to grow from seed but are also readily available as starts for those who prefer to buy plants and for those who live in colder climates with a short growing season.

direct sow

- Bush beans
- Radishes

direct sow or use starts

- Chard
- Collards
- Cucumber
- Dill
- Kale
- Leaf lettuce
- Okra
- Summer squash
- Zucchini

use starts

- Basil
- Parsley
- Peppers
- Tomatoes
- Eggplant

SOME PLANTS LIKE IT COOL

Plants thrive at different temperatures, so plan to grow crops during the seasons they prefer. Cool-season plants such as kale, collards, broccoli, cauliflower, cabbage, chard, carrots, lettuces, and beets require cooler temperatures and should be planted in late summer for a fall harvest or in early spring as soon as the ground has thawed and the soil can be worked. (Not all these plants are recommended for a beginner, but I've listed them because you'll need this information to complete later stages.)

These cool-season plants fail to perform when temperatures get too high. In fact, cool-season plants become stressed and "bolt" when

Leafy greens like lettuces are a great beginner plant because they're relatively easy to grow and are one of the more costly items in the grocery store.

temperatures start to rise—they stop producing foliage and instead begin to flower and set seed. This is the plant's last-ditch effort to send its off-spring into the world before it dies. Once bolting begins, there's no stopping it.

Cool-season plants can withstand some level of frost. In fact, some of them are so frost tolerant that in mild climates they're able to survive winters with little to no protection. Here are some cool-season crops, listed in order from most to least frost tolerant (frost tolerance can be further enhanced by selecting varieties suited for overwintering).

AND SOME LIKE IT HOT

Warm-season plants, as you can no doubt guess, are those that thrive in warmer temperatures and cannot reliably withstand any level of frost. I was once lucky enough to have a frostbitten flat of tomato seedlings miraculously survive, but I would never intentionally subject my warm-season plants to frost. These heat-loving plants are placed in the ground once all chance of frost has passed, otherwise you risk losing them all to a cold snap. Furthermore, seeds of heat-loving plants like tomatoes, peppers, and beans won't even germinate when soil temperatures are too low.

cool-season plants

- Kale
- Collards
- Leaf lettuces
- Parsley
- Chard
- Radishes
- Broccoli
- Cabbage
- Dill

warm-season plants

- Basil
- Bush beans
- Cucumber
- Eggplant
- Okra (warmer climates only)
- Peppers
- Summer squash
- Tomatoes
- Zucchini

Bush beans like this yellow wax bean are some of the easiest seeds to direct sow.

It's easy to add a few pollenizer plants to your food garden. Try marigolds, snapdragons, or cosmos (from starts) or direct sow a handful of sunflower seeds.

FIND YOUR IDEAL PLANTING DATES

The exact timing of your spring and summer plantings depends on where you live. In general, American gardeners at more northerly latitudes have shorter, cooler growing seasons compared to those growing farther south. You'll need two key pieces of information to guide your decision on when to plant what: (1) the average dates for the first fall frost and last spring frost in your region and (2) a planting chart. Both are available from the National Gardening Association, which allows you to enter your zip code to get the relevant information.

FROST-FREE DATES

In my region, the average last spring frost and first fall frost are May 5 and October 9, respectively. These dates represent the average frost-free period in which I can most reliably plant and grow warm-season plants. This means I shouldn't plant warm-season starts until after May 5, when the chance of frost has passed for my area. My goal is to harvest from warm-season plants before the killing frost comes in autumn. I can sometimes stretch this range a bit in either direction by keeping an eye on the long-term weather forecast and gathering information from local growers, but I use these averages as my starting point.

Having a limited window for growing is another good reason to plant starts instead of seeds for certain crops. For example, tomatoes don't begin producing fruits until late in summer. I'd get about three tomatoes if I were to plant seeds at the beginning of May. Planting starts gives me a month or two head start on the season. By comparison, the plants I recommend on page 27 for direct sowing require a shorter amount of time to reach maturity and thus are safe bets to start from seed immediately after your last spring frost date.

USEFUL GARDEN TOOLS

- Digging fork for aerating soil, harvesting root crops, and removing plants
- Garden rake for spreading soil, compost, or mulch
- Shovel for digging large holes or moving soil
- 10-tine mulch and compost fork for shoveling mulch and compost
- Hand trowel for planting starts and heavy weeding
- Clippers or heavy scissors for harvesting vegetables and flowers
- Bypass pruning shears for cutting large stems and twigs
- Wheelbarrow for transporting heavy or bulky items

PLANTING SUGGESTIONS FOR 4 × 8 BEDS

Try this garden plan for two 4 × 8-foot beds. The lettuce and radishes may appear to be planted too close to the tomatoes, but these early crops will be harvested long before the tomatoes are fully mature. Okra was not included as it is grown only in warmer climates. If you sub in okra, follow the spacing recommendations in Appendix B. Following spacing recommendations encourages proper plant growth and development and prevents disease.

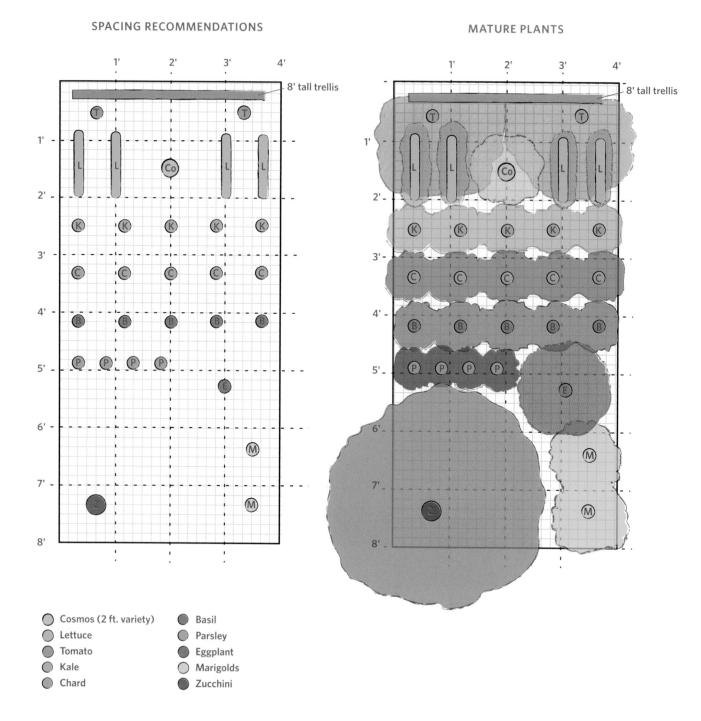

SPACING RECOMMENDATIONS

MATURE PLANTS

8' tall trellis

- ◯ Cosmos (2 ft. variety)
- ◯ Lettuce
- ◯ Tomato
- ◯ Kale
- ◯ Chard
- ◯ Basil
- ◯ Parsley
- ◯ Eggplant
- ◯ Marigolds
- ◯ Zucchini

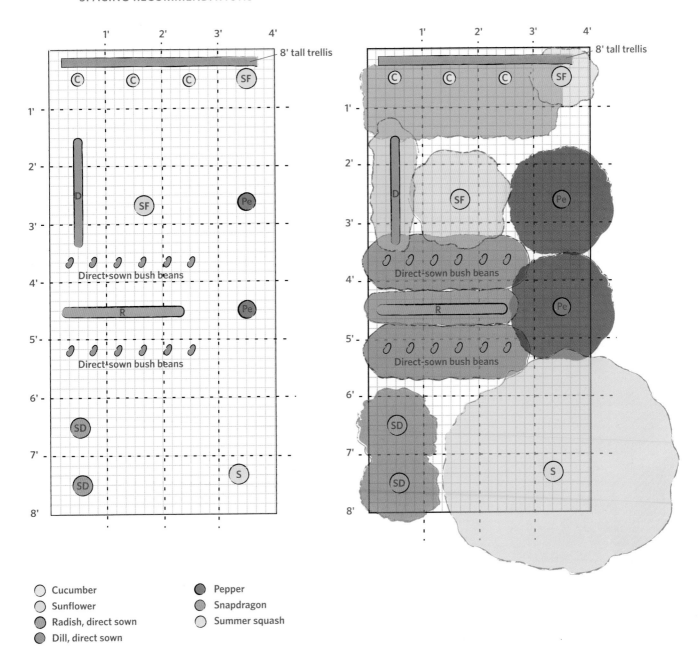

SPACING RECOMMENDATIONS

MATURE PLANTS

8' tall trellis

○ Cucumber
○ Sunflower
● Radish, direct sown
● Dill, direct sown
● Pepper
● Snapdragon
○ Summer squash

Direct-sown bush beans

GET OUT AND GROW!

Now that you've built your small beds and have planned what you're going to plant and when, it's time to start growing! Don't expect everything to go perfectly; as with most things in life, gardening is a process. Enjoy yourself, have fun with the different steps, and recognize that gardening is a lifelong learning experience. You'll have new challenges and successes every year that will hone your expertise and build your confidence in the garden.

DON'T CROWD CROPS

Most gardeners, myself included, have a tendency to pack too many plants in a bed. This is all fine and good when the plants are small, but their mature versions won't develop properly if not given adequate space. Planning out the beds therefore requires some knowledge of mature plant sizes.

Lettuce and parsley, for example, require a circular area with a diameter of about 6 inches. Kale, chard, and snapdragons need 10 inches. Collards need 16 inches. Cucumbers, peppers, and tomatoes need 2 feet. Zucchini and summer squash need 3 feet. Sunflowers and marigolds are highly variable, as some varieties remain small while others can grow to be quite large. Before planting, be sure to check the plant tag or seed packet for each variety's requirements.

Some people use twine to divide their beds into evenly spaced squares; these guides help prevent the overcrowding of plants. You can accomplish something similar, and more exact, using dinner plates of varying sizes or circles cut from cardboard.

Remove the tips of young seedlings such as basil, marigolds, and cosmos (shown here) to encourage bushier, fuller growth.

Cardboard circles of different sizes allow you to plant your garden taking into account the mature size of various vegetables and flowers.

Planting Starts

1 Water the starts well before planting. Use a trowel to dig a hole 2 inches wider than a start's rootball and an inch or two deeper than the planting depth for that particular start. Place a small handful of compost or 1 to 2 teaspoons of granular organic fertilizer such as Espoma Plant-tone in the bottom of the hole. I most often brush a bit of my top-dressed compost into the hole before planting to save time.

2 Lightly massage apart the roots of the start if it appears rootbound.

3 Place the rootball in the hole and lightly pack the soil in place around it. The soil should have good contact with the rootball to encourage root expansion.

4 Water the soil well and continue to water daily for the first three or four days if you don't receive ½ inch or more of rain per day. After that, monitor your soil moisture, and water every few days if it isn't raining.

Seed Planting

1   Use your finger or a small stick to create a shallow trench with a depth that's suitable for the particular seed size. Larger seeds like zucchini or sunflower can be pushed directly into the soil without a trench.

2   Sprinkle the appropriate number of seeds into the trench following spacing recommendations on the seed packet or in Appendix A.

3   Brush the soil back into place and lightly pat the surface to help keep the seeds in place and prevent the soil from washing away. Gently water the newly planted seeds to settle them in (see How to Water Effectively, page 38).

PLANT AT THE RIGHT DEPTH

Plants fall into one of two categories—those that can be planted deeper in the soil than the start's existing soil line and those that need to be planted at the same depth. Tomatoes, peppers, tomatillos, marigolds, and a few other plants can be planted deeply because their stems produce roots once underground. Burying the stems of these plants to just below the first set of leaves improves root development, helping to establish a strong, healthy plant. Brassicas (kale, collards, cabbage, broccoli, cauliflower, and others) also benefit from being buried a bit deeper because the soil can support their stem and keep them from flopping over.

The majority of plants—including lettuces, okra, basil, dill, parsley, and beans—are incapable of growing roots from their stems. Burying these plants deeper than their existing soil line will cause the plants to rot.

ADDING SEEDS AROUND PLANTED STARTS

Read the seed packets. Most packets are printed with the basic information you need to plant seeds—including row spacing, planting depth, planting density, thinning recommendations, and sometimes ideal soil temperatures. Seed packets for greens such as kale, lettuces, and chard also provide planting instructions and spacing requirements for growing baby as well as full-size leaves.

Measure and mark sowings. Unlike starts, seeds take days to weeks to make their aboveground appearance, and sowings should therefore be marked using sticks, twine, or plant labels. Unmarked sowings can be easily forgotten or accidentally sown too closely to another plant. Use a tape measure to determine planting distance from other seeds or starts (measure from the center of the start). Many experienced growers measure in this way as it's a foolproof method for ensuring proper spacing.

Check for soil moisture before planting. Lightly water the soil before planting, if necessary. Planting in moist soils (not muddy or soaked) means you don't have to water the seedbed for as long after planting and thus lowers the risk of washing newly planted seeds away.

Plant at the right depth. The general rule is that seeds should be planted at a depth that equals two to three times the length of the seed. Smaller seeds such as radish, lettuce, kale, collard, and dill should be planted about ¼ inch

The first set of leaves to emerge from a seedling—the so-called seed leaves—are not always shaped like the leaves of the mature plant. The second set—known as the true leaves—and every set thereafter take on the mature plant's shape, as shown by this tiny kale seedling.

deep. Medium-size seeds such as zucchini, summer squash, and cucumber can be planted ½ to 1 inch deep. The largest seeds, such as bean and sunflower, should be planted at a depth of 1 inch.

Plant in rows or mounds. Most seeds (and plants for that matter) are planted in rows to ensure sufficient spacing and uniform acquisition of nutrients and water through properly developed roots. The exceptions are plants like summer squash and zucchini, which are planted in mounds to accommodate their bushy habit and because they grow far too large for a traditional row. You'll likely only grow one or two of these

When working with a small amount of space, place large or sprawling plants like zucchini and squash in the corner of a bed and train the plant to grow outside the bed to save space.

PLANTING TERMINOLOGY

>>> **Planting depth.** The depth at which a seed should be placed in the soil. Planting seeds too shallowly can prevent proper root establishment, whereas planting them too deep can prevent germination altogether.

Soil line. The line where a direct-sown plant transitions from roots to stem. This transition line should be maintained when planting most starts: That is, don't bury any part of the stem. Some plants, like tomatoes, will grow roots from the stem; these plants benefit from being planted more deeply. (See Plant at the Right Depth, page 35.)

Seed spacing. The amount of space between newly planted seeds. Seed spacing is represented in inches between seeds or number of seeds per foot. Do your best to follow the recommended spacing, but know that you can always thin plants later. Lettuce seeds, for example, are difficult to space appropriately as they're so small.

Plant spacing. The amount of space required between mature plants to ensure proper growth. Seeds are often planted at higher than optimal densities to account for loss during germination. Thin seedlings to the indicated plant spacing once they're well established.

Row spacing. Plants are most often planted in rows. Allowing adequate space between rows will encourage proper plant development and prevent disease.

left: Seedlings should be thinned once they're well established to allow space for proper growth and development.

right: When thinning, follow the spacing recommendations on the seed packet or refer to Appendix A for guidance.

squash plants at a time because they produce so abundantly.

Water well. Newly planted seeds need more attention than starts when it comes to watering because seeds simply won't germinate if not kept consistently moist. Most seeds germinate within a week, but some, like carrots, can take up to three weeks. (Because of their finicky germination requirements, I don't recommend carrots for beginners.) Water newly planted seeds daily, preferably in the morning, until the plants have their first set of true leaves (see photo on page 35).

Watch the weather. A few days of gentle showers will greatly improve germination success and spare you from having to constantly water. However, avoid planting before heavy rainfall, otherwise small seeds can wash away or seeds or seedlings can rot in place.

Thin established seedlings. In many cases, you'll thin seedlings once they're well established to give them the room they need to grow well. It's tempting to thin them immediately after they've emerged from the ground,

but it's best to wait two or three weeks and thin when the seedlings are about 2 inches tall. You may lose a few early seedlings due to over- or under-watering or damage from pests, so wait until you're certain you have a good survival rate before thinning.

Don't wait too long to thin. Waiting too long to thin can be equally as problematic as thinning prematurely, especially if the seedlings are crowded. Plants establish strong, healthy roots and stems during their early developmental stages. Overcrowded plants that compete for space and resources often become weak and spindly as they try to "outreach" their competitors.

Radishes, for example, never develop into spicy little orbs if their roots aren't provided with the space to do so. As plants become larger, overcrowded living conditions not only prevent proper growth but also encourage the spread of fungal and bacterial diseases. Proper airflow and space are essential requirements for plant growth, and thinning is therefore nonnegotiable regardless of how destructive it may feel.

Water the garden with a hose nozzle set to a gentle shower, especially after you've just planted seeds. Keep the stream moving so that water doesn't pool on the soil surface and accidentally wash away the seeds.

A soaker hose is a good option for smaller gardens. Large spaces require extremely high water pressure to evenly disperse water the entire length of the soaker hose.

HOW TO WATER EFFECTIVELY

>>> Over- and under-watering are two of the biggest gardening mistakes. Plants aren't nearly as finicky once they're well established in the garden, but giving them adequate water (not too much, not too little) is imperative in the beginning. The roots of young plants take time to grow into the surrounding soil. Water availability is therefore limited to what exists in their tiny rootball. If the soil is dry around them, it will wick moisture away.

A plant's roots will begin to grow into the surrounding soil after a few days, but until then, these tiny balls of soil can quickly become depleted of water when sitting in the sun for a day, even when the surrounding soil appears to be moist. If possible, time plantings around predicted rain in the forecast, to reduce the need for hand watering.

Be gentle. Use a hose nozzle that gently showers the soil, mimicking rainfall. The goal is to soak the soil, not flood it or wash away the top layers. If using a nozzle, keep it moving; if you spray in one spot for too long, you'll wash away the soil or seeds. This is especially problematic for newly planted seeds. A soaker hose is a good option for small spaces. Install it along the row before planting.

Water deeply. Watering two 4 × 8-foot beds should take about 10 minutes when using a hose and nozzle, 40 minutes when using a soaker hose. Check to see if you've watered enough by poking your finger a few inches into the soil. Is the soil moist at that depth? If yes, you're done! If no, keep watering! Soaking into deep layers of soil often takes more time than you'd expect.

Study the soil. Have you ever noticed that even when you're constantly adjusting your aim, water starts to pool on the soil surface? This usually means one of two things: Either you've saturated the soil and the watering session is complete, or the soil is so dry (or full of clay) that it can't absorb the water at the rate you're applying it. If after a watering session the soil is dry when you poke a finger in, water again for 10 minutes with less flow, walk away for 15 to 20 minutes, then come back and water the area yet again. By allowing time to pass between gentle watering sessions, you help the soil better absorb moisture. Look into soaker hoses if dryness is an ongoing problem.

HOW TO SUPPORT TALL OR VINING PLANTS

Vining plants that are trained to grow vertically rather than along the ground are less prone to disease and insect infestation; they are kept drier thanks to improved airflow and are more difficult for crawling insects to reach. Cucumbers, for example, don't necessarily have to be grown vertically, but doing so can improve production and prevent insect damage. Vertical growth is also a major space saver, as one or two unsupported cucumber vines can easily sprawl over an entire 4 × 8-foot bed.

Cucumbers, pole beans, peas, and many other crops will naturally climb with little to no assistance. Simply provide them with adequate support and they'll tendril their way up. Tomatoes and peppers, on the other hand, must be guided by attaching their stems to a support using garden twine. Be sure to install stakes and trellises before planting, because their insertion into the ground can damage immature plant roots.

I recommend staking peppers and tomatoes and trellising cucumbers. Different varieties of these plants grow to different heights, and this will influence the trellis or stake design. Read the variety's description before building a trellis or select a variety suitable for the trellis you've already built. You may also need to stake sunflowers, especially if you're growing a top-heavy, single-flowered variety.

BUILDING A CATTLE PANEL TRELLIS

Bend a 16-foot cattle panel over a walkway and wedge it between two T-posts or other sturdy anchors. Attach the panel to the anchors using zip ties or wire. Sow seeds or place starts at the base of the arbor and train the plants to grow up and over the structure. A panel like this could be bent between two 4 × 8-foot beds (with the walkway beneath the trellis) and easily accommodate indeterminate tomatoes (see page 41) and cucumbers.

Cucumber varieties like this 'Suyo Long' can vine up to 10 feet. Know the variety's mature size and build a trellis accordingly. Select compact varieties if you have limited space.

RECOMMENDED STAKES AND TRELLISES

My personal collection of stakes and trellises meets four important criteria. First, supports must keep plants upright for the duration of the season. Under the weight of full-grown plants, weak and inadequate supports can topple, which is a massive pain to correct. Second, supports must be affordable. Third, I have minimal carpentry skills and thus supports must be easy to construct. Last, supports must be mobile so they can be moved about the garden, because I avoid planting cucumbers, tomatoes, and peppers in the same location every year.

Steel stakes with plastic coating. These lightweight stakes are available in varying lengths and thickness and work best with flowers and other plants that require minimal support. Alternatively, you could employ a tall piece of bamboo or any long, straight stick. (Do not use freshly cut bamboo; it can take root and potentially become invasive.)

Metal fence posts. These are some of the handiest supports because they're sturdy and versatile, and they last for years. Driving a T-post (or U-post) into the ground takes a bit more time than simply placing a stake, but it's time worth investing. Five-foot-tall T-posts are my preferred stake for peppers and other short-statured plants that produce heavy fruits. These posts also serve as excellent anchors to which trellis materials or finicky cages can be attached.

Cattle panels. Cattle panels (a.k.a. hog panels) attached directly to T-posts are among the easiest options to install. Eight-foot panels can be secured in an upright position, and 16-foot panels can be bent between T-posts for use as arbors.

Tomato cages. Tomato cages are popular because they're easy to install but they have few applications as they're prone to toppling over. Cages can be used for peppers or determinate tomatoes, but they'll need to be secured to a T-post or other sturdy stake to prevent them from falling over. Cages with wide bases that taper toward the top are slightly sturdier than models that taper at the bottom.

A 16-foot cattle panel bent between two 4 × 8-foot beds (with the walkway beneath the trellis) accommodates indeterminate tomatoes and cucumbers.

INDETERMINATE

DETERMINATE

TWO TYPES OF TOMATOES

>>> Tomatoes come in two varieties: indeterminate or determinate. How a plant grows and is supported depends on its type, so it's important to understand the distinction.

Indeterminate. These tomatoes can vine indefinitely; they continue to send out flowering stalks and produce fruit over the course of the summer. Without support, indeterminate tomato plants sprawl over the ground. I've had good success training a single stem of each plant up a stake and regularly removing any suckers that appear. Suckers are offshoots that, if left in place, will mature into large stems. Their removal maintains a tidy plant and improves air circulation, thus preventing disease.

Determinate. These varieties have a predetermined height at which they'll stop growing. Unlike indeterminate varieties, determinate tomatoes generally produce one large, single crop. They have a bushier habit and are bred to be less than 4 feet tall. You'll still need to prune them to improve airflow, but you can take a far less aggressive approach. Allow the plants to develop four or five stems, rather than just one or two. Determinate tomatoes are best grown in a tomato cage rather than supported by a single stake.

Staking Single-Stem Tomatoes

Tomatoes can easily grow into unruly, disease-prone vines if not properly supported. For indeterminate tomatoes, I use the single-stem staking method to provide each plant with adequate airflow, thus preventing blight (a soilborne fungal infection) and other diseases. Tomatoes often succumb to blight by the end of the season regardless of the staking method, but proper preventive measures allow for a successful, productive plant even if the leaves become infected.

1 Drive a 4- to 5-foot-tall T-post into the ground to act as an anchor. Secure an 8- to 10-foot-tall stake to the post.

2 Dig a hole right next to the stake. Remove the bottom leaves from a tomato start and plant it, burying 6 to 8 inches of the stem. Mulch around the plant to avoid having soil splash up onto the leaves, potentially spreading blight.

3 Every 10 to 14 days, remove suckers with clean pruning shears or sharp scissors and loosely tie the plant to the stake with garden twine.

4 As the plant grows, remove leaves that touch the ground or appear to be dying. Discard the leaves rather than composting them, because they can potentially spread soilborne disease.

Flowering stalk—a stalk that originates on the internode, not from a node, and eventually produces flowers

Bud—the structure that grows into a new stem

Sucker— a bud allowed to develop into a side shoot

Internode—the space on the stem between nodes

Node—the place where the stem and a leaf meet; the point of bud origination

Fruit

Seeds

START COMPOSTING

Making your own compost is a great way to reuse spent plant materials from the garden and yard while keeping food scraps out of the landfill. Composting can be as haphazard or meticulous as you'd like, but you'll be able to produce compost quicker if you achieve the correct ratio of green and brown materials, keep the pile aerated, and ensure that the pile remains moist.

THE THREE-BAY SYSTEM

I use a basic three-bay compost bin constructed from wooden pallets tied together with wire. There are certainly more aesthetically pleasing designs, but this construction takes almost no time and costs virtually nothing. You can get away with having just one or two bays, but building three bays allows one or two piles to be actively decomposing while the third serves as the recipient of new materials.

When one heap matures, spread its contents around the garden, then use a pitchfork to transfer a younger heap into the newly vacant space. This process aerates and turns the compost, which ultimately speeds up decomposition.

Moving heaps from one bin to another is also a good opportunity to layer in additional ingredients if the compost is lacking.

GETTING THE RATIO RIGHT

Additions to a compost heap are broadly categorized as either green or brown. Green materials are generally living or have recently been harvested (such as kitchen scraps or weeds pulled from the garden) and include fresh animal manures. Green matter is high in nitrogen. Brown materials are generally dried or lack greenery—things like sticks, dried leaves, and paper scraps. Brown matter is a source of carbon.

Some experts recommend a 20:1 ratio of brown to green ingredients, but I simply use whatever green materials I have on hand, and I amend the heap with browns as I go. I use my senses (primarily smell) to determine if I've added too much of one or the other material. A properly constructed compost heap should smell earthy and plantlike, not putrid.

IS IT COMPOST YET?

>>> The rate at which compost decomposes depends on several factors, including temperature, moisture, and the ingredients you've added. I expect my compost heaps to take about a year to mature, given my lackadaisical approach to carbon and nitrogen ratios, but generating a healthy compost can take as few as two to three months if you're more focused on your ratios, moisture levels, and aeration.

I live in a region with regular precipitation, so I don't have to water my compost heap. But if you live in an arid climate, watering may be necessary to keep the microbes alive and actively breaking down the materials you add. In those cases, you may want to think about locating the compost heap where it can easily be watered.

A three-bay composting system is easy to maintain and provides plenty of compost.

Too much brown material. If you find that the pile is decomposing much slower than expected, you may have added too many dried leaves or other brown materials. Without sufficient nitrogen from green materials, the carbon sources degrade very slowly. The solution? Simply incorporate more greens.

Too much green material. Carbon performs the same function in a compost heap as it does for the soil: It provides a place to store nutrients such as nitrogen. Adding green material without enough brown results in excess nitrogen being off-gassed, as it has nowhere to be stored. This off-gassing is one reason why a compost pile may stink. On the positive side, a pile with a lot of nitrogen generates more heat, which can help kill weed seeds and speed up decomposition.

Too much water. A compost heap starts to stink if it becomes overly saturated with water. Adding too much high-moisture green matter and not enough dry brown material is one reason for this; weather patterns can also be a factor. Some gardeners cover their compost with a waterproof barrier to have more control over the moisture content.

If you find yourself with a stinky compost heap, simply turn the compost into an empty bin, layering in plenty of brown material between scoops. Cardboard scraps work particularly well for this application because they absorb moisture. Try to break up any saturated clumps of compost while transferring the heap.

MATERIALS TO ADD (OR AVOID)

There are many suitable additions to a compost heap, but it's also important to note that certain items should *not* be added. Eggshells are okay, for instance, but meat and other animal products are no-no's because they can attract unwanted critters and take a long time to decompose. Discard diseased plants away from the garden and compost pile, as many diseases can persist in the soil for years.

Cardboard, paper plates, and paper scraps are excellent compost additions, but be mindful of paper goods with dyes and the addition of synthetic materials such as plastic or glossy coatings, tape, or glassine envelopes. Be aware that disposable cups, plates, and cutlery made from plastic-like biodegradable plant materials are unlikely to decompose in a home composting system. These products are designed to decompose under high temperatures that are only achieved in industrial composting systems.

brown materials (carbon)

- Paper products (nonglossy)
- Corrugated cardboard
- Wood chips or wood shavings
- Small twigs
- Sawdust
- Dried leaves
- Dried, pesticide-free grass clippings

green materials (nitrogen)

- Fresh, pesticide-free grass clippings
- Coffee grounds
- Fruit and vegetable scraps
- Spent plants and weeds from the garden
- Eggshells
- Animal manures

ADD COMPOST ANNUALLY

Annually top-dressing your beds with compost is absolutely necessary, especially when you consider that you're harvesting many of the nutrients in the soil when you harvest your crop. Many gardeners assume that they can add compost in small quantities every now and again, but I favor a more aggressive approach.

For older, well-established beds that produce abundantly and don't appear to require huge nutrient inputs, I top-dress with an inch or so of compost or aged manure every spring. Sometimes I even skip a year. For younger beds or those that aren't performing well, I top-dress with 2 inches of compost every spring and sometimes again in fall.

Regularly top-dressing newer beds with compost gets them off to a good start.

WEEKLY GARDEN MAINTENANCE

A garden's longevity, production, and overall health improve when the site is well maintained. And regular maintenance saves you time in the long run. For example, pulling weeds before they go to seed prevents them from reseeding, and proactively removing diseased plants limits disease spread. These simple tasks require a minimal time investment when dealt with promptly, whereas procrastination yields more labor-intensive chores.

To tend two small vegetable beds requires approximately one hour of maintenance per week. Planting and cleanup weeks may entail an additional one to two hours, while some weeks during the growing season may call for no maintenance at all! Whatever the case, survey the garden frequently and carve out some time each week to address any needs. The following tasks will help you get started.

- Remove spent plants by clipping them off at the soil line and tossing them in the compost pile. Leave the roots underground to help feed the soil.
- Replace spent plants with seasonally appropriate substitutes.
- Remove bolting plants unless you're hoping they reseed for next year.
- Deadhead (remove spent flowers) to encourage continuous blooms throughout the season.
- Ensure that vertically grown plants are properly climbing and supported by stakes.
- Keep a close eye on the weather forecast to anticipate when you'll need to water. Watering in the morning is best but not always practical.
- Mulch exposed soil for protection, weed prevention, and moisture retention.
- Pull the weeds, roots and all. Don't clip them at the soil's surface or they may regrow.

HARVEST THE BOUNTY

Having the opportunity to harvest your own homegrown fruits and vegetables is one of the greatest joys of gardening. As with maintenance, be sure to stay on top of the harvest so that you don't accidentally miss a peak window of ripeness.

HARVESTING LEAF CROPS

Leaves are some of the first edibles to mature but can become tough and bitter when too large. Harvest leafy greens such as kale, collards, and chard when they're tender and appropriately sized (12 to 15 inches in length from the tip of the stem to the end of the leaf for kale and chard,

and about 18 inches for collards; check the cultivation instructions for each specific variety for more guidance). Many leafy greens—including kale, collards, chard, dill, lettuce, and parsley—are harvested in what's known as a cut-and-come-again fashion. The lowest or outermost leaves are removed, and the plant is left in place to continue producing.

Leaf lettuces can either be cut off 1 to 2 inches above the soil line and allowed to regrow or mature outer leaves can be removed individually. Head lettuces such as romaine are harvested in their entirety once fully mature. Lettuces are quick to produce compared to heartier leaves like

Grasp the entire head of lettuce and use scissors to remove the leaves 1 to 2 inches above the soil line. The plant will regrow and be ready to harvest again in about two weeks.

Outermost leaves can be pinched off individually as they mature. This technique is more time-consuming but results in a tidier, more uniform harvest.

Harvest lower, mature leaves from kale, collards, chard, and other leafy greens. Continue to do so until the plant starts to shoot out a flowering stalk and stops producing leaves.

WASHING AND STORING GREENS

>>> Mulched beds keep your greens cleaner, but regardless, you'll need to give greens a quick rinse before you eat them. Fill the sink with water, submerge the greens, then gently swish them about to remove dirt or other particles. To dry kale, collards, chard, and herb bundles, simply gather the rinsed bunches by their stems, step outdoors, and swirl the greens through the air to shake off the water. To dry leaf lettuces, you can either use a salad spinner or place the leaves in the center of a clean kitchen towel, gather the four corners, step outdoors, and swing the bundle around. Fluff the bundle and repeat.

Washed and dried greens can be stored in the refrigerator in a sealed bag or container. Greens that are too large to fit into a bag can be chopped before storage. Kale and collards keep for as long as 2 weeks (1 week, if chopped); lettuces and chard usually last about 10 days. You'll be surprised by the longevity of your freshly harvested greens compared to those purchased at a grocery store.

kale and collards, but they have a narrow window of production. Leaf lettuces can usually be cut and harvested three to five times before they start to bolt or become bitter.

Once lettuces start to bolt (grow upward and produce a flowering stalk), you'll notice the presence of bitter undertones caused by a milky substance released from the veins where the leaf was cut. The substance isn't harmful, but its bitter flavor is less than ideal, and it indicates that your days of bountiful lettuce are coming to an end.

Because warmer temperatures encourage bolting, lettuce becomes more difficult to grow during warmer months. The same is true for kale, chard, and collards, although these tend to be more heat tolerant than lettuce.

Basil is harvested differently. For a healthier, more productive plant, harvest entire stems directly above nodes and even pinch off new whorls of leaves that grow from the top. Buds will grow out from the nodes and form new stems, improving production.

HARVESTING ROOT CROPS

Radishes are the only root crop that I suggest for a beginner grower. Small salad radishes mature in as little as four weeks, and you'll want to keep your eye on the crop so that it's harvested on time. To check for readiness, simply place your thumb and pointer finger below the surface on either side of the radish root to see if it's an appropriate size (look at the seed packet to determine the variety's mature size). If yes, it's time to harvest! If not, push some dirt back in place around the radish root and check again the next week. Radishes that are left in the ground too long can develop a tough outer skin and begin to crack.

Washing the roots immediately after harvest is best, as dirt that dries in place becomes more difficult to remove. Many root vegetables have edible greens, some with strong flavors that are often delicious additions to soups or salads. If you plan to eat the greens, store them separately

above: Rather than remove individual leaves from a basil plant, cut off entire stems directly above nodes to encourage the plant to bush, thus producing a better yield.

below: The upper part of a radish root is sometimes visible above the soil line, making it easy to judge readiness.

from the roots, because roots will store longer (one to three months). As with leaf crops, roots are best stored in sealed containers or bags in the refrigerator.

HARVESTING OTHER VEGETABLES

Many vegetables—including zucchini, cucumbers, summer squash, bush beans, and okra—are eaten long before the fruits (yes, these are fruits!) are fully mature. Okra, for example, is harvested when the fruits are 4 to 5 inches long. Miss this optimal window and you'll find yourself gnawing on a leathery snack that's far too fibrous to enjoy. Likewise, zucchini and summer squash can grow extremely large and develop a tough outer skin, at which point the flesh becomes watery and full of large seeds. These transformations from tender to tough take mere days, so it's best to check fruits daily and to harvest when they're young.

Tomatoes and peppers are harvested once fully ripe and mature. Green peppers will always ripen further into a color (red, yellow, orange, or purple), but they can be harvested when green, if desired. Tomatoes can also be harvested when green, placed indoors (preferably in a sunny window), and allowed to ripen. Waiting for peppers and tomatoes to ripen on the vine will yield the most delicious fruits you've ever tasted, but a premature harvest is occasionally necessary. Tomatoes on blight-infested plants can sometimes be saved by early harvesting. And you'll need to harvest all tomatoes and peppers, regardless of their developmental stage, immediately before a pending frost.

Female flowers are distinguishable by the presence of a tiny fruit at the base of the flower.

UNDERSTANDING FRUIT FORMATION

>>> Understanding a plant's life cycle is a key indicator of fruit formation. Once a plant blooms, look for pollinating insects and then for developing fruits. Squash, zucchini, and cucumbers have separate male and female flowers that are distinguished by the presence of a miniature fruit at the base of the female flower. Male flowers are often more abundant, especially earlier in the season, but you can anticipate fruit development once you see female flowers in bloom. Okra, beans, tomatoes, and peppers have male and female parts on the same flower. If pollination is complete, most flowers will mature into fruits.

S T E P

2

FOUR HUNDRED

square feet

Now that you've mastered one or two small beds, you're ready to expand your garden! Increasing your cultivation space to 400 square feet allows for greater production and opportunities to grow new plants. As you'll learn in this chapter, increased diversity has a range of benefits that improve your garden's productivity while supplying you with a greater abundance of food.

Place small, quick-to-mature plants like lettuces between slow-growing plants like cauliflower. As the lettuces are harvested, space opens up for the developing cauliflower.

the plan

- **ADD MORE GROWING AREA.** No-dig beds can be created from even larger pieces of organic matter than you may have imagined possible. Hügelkultur beds (see page 58) are built much like sheet-mulched beds, but the inner layers are made from logs. These beds are especially good at holding water and providing your plants with long-lasting nutrition.

- **DIVERSIFY YOUR PLANTINGS.** With more space, you'll be able to add a few perennials, including herbs and flowers. You can also start experimenting with carbohydrate-rich vegetables such as potatoes, beets, carrots, and turnips.

- **DEVELOP YOUR PLANTING SKILLS.** Start seeds indoors to get a jump on the season and mix in succession crops to maximize production.

- **LEARN ABOUT SOIL AERATION AND MULCHING.** Continue to improve and protect your garden soil to create a sustainable environment for pest management, disease suppression, and water retention.

YEAR 3: EDIBLE PERENNIAL BEDS WITH WIDE
WALKWAYS TO PREVENT SPREAD

YEAR 4:
MIXED
ANNUAL AND
PERENNIAL
BEDS

YEAR 1: MAIN GARDEN:
NARROW WALKWAYS AND
WIDE BEDS TO OPTIMIZE
SPACE

YEAR 2:
BLACKBERRY
ADDITION
WITH WIDE
WALKWAYS TO
ACCOMMODATE
SPRAWL

I've added new
beds to the
perimeter of my
garden nearly
every year and
have designed
the bed and
walkway widths
based on my
intended use.

BEFORE YOU DIG IN

Determine where to place your new beds using the same site selection criteria from Step 1. If you plan to construct raised beds, calculate the expense before starting, as this large expansion can prove costly. Creating a few more borderless, in-ground beds with sheet mulching will be more affordable. You may want to explore a new construction technique: hügelkultur beds.

Gardening is not prescriptive; it's a fluid process that demands creative problem-solving for each situation and challenge. There's no right or wrong way to design your garden, but there are a few considerations to keep in mind. Gardens should be both aesthetically pleasing and functional, which means you'll need to balance the two when thinking about your final design.

Choose appropriate walkway widths.
Wide walkways (3 to 4 feet) are easier to navigate and provide overflow space for larger plants but demand more space and maintenance. Narrower walkways (18 inches) are more difficult to navigate and can be overtaken by unruly plants but require far less space and maintenance. Consider space limitations and your physical ability to maneuver tight spaces in a garden when choosing walkway width.

DESIGN YOUR LAYOUT

Each of these design ideas offers comparable growing area but with a different footprint. Design A utilizes wide beds and narrow walkways for growers needing to optimize space. Design B widens the walkways around narrower beds for increased maneuverability. Design C creates an aesthetically pleasing garden to accent a home or other building.

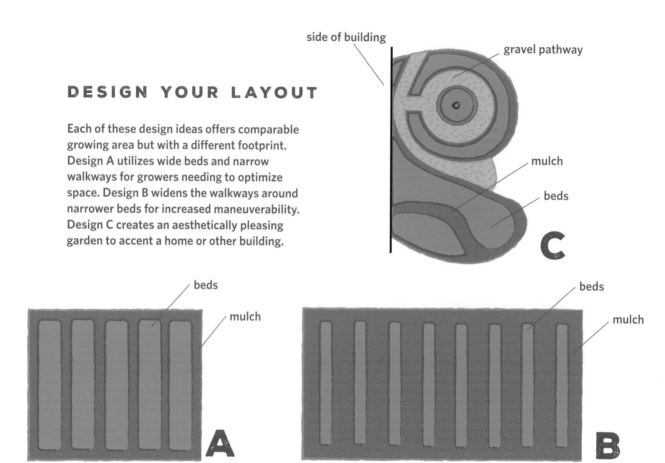

side of building

gravel pathway

mulch

beds

beds

mulch

C

beds

mulch

A

B

Don't make beds too wide. Wide beds (3 to 4 feet) optimize space utilization while narrower beds (2 feet) are easier to work with. Whatever the width, you need to be able to easily reach the center of the bed. Measure a template on the ground and practice reaching to all corners of the shape to ensure that it's an appropriate size.

Create functional shapes. Beds can be arranged in a rectangular pattern with even, parallel walkways, or unique geometric shapes that give a garden more aesthetic appeal. I use both techniques but prefer the long, rectangular beds that are no more than 3 feet wide because they're better suited for winter gardening (see page 139).

Think about future expansions. Choose a layout that can be expanded not just this year but in coming years. Deciding the exact placement of future beds can be challenging, but you'll limit yourself if you choose a complicated, standalone design.

MANAGE PESTS SUSTAINABLY

Proper nutrition prevents both chronic and acute illness in humans. Plants are no different. When provided with adequate nutrition through the soil, plants are better equipped to protect themselves from insects and disease. Building healthy soil takes time, but if you top-dress your soil with compost annually, you can expect fewer problems in the long run.

Planting flowers and a wider diversity of food plants provides a second line of defense against insect pressures. Flowers attract pollinators, of course, but also bring in predatory insects and birds that can feast upon problematic caterpillars, grubs, moths, and so on. When you plant flowers, you're enhancing your diversity and essentially building a protective food web. I aim to plant 30 to 35 percent of my garden with flowers and do so by using a mix of perennials and annuals. Perennials bloom year after year, which saves time, money, and energy, while annuals tend to bloom for longer stretches of the growing season.

A sustainable pest-management approach focuses on improving soil health, diversifying plantings, physically removing pests, and utilizing physical barriers recommended in later chapters. Avoid pesticides as much as possible, with the goal of eliminating them entirely. Pesticides, including those that are certified organic, have ecological consequences that disturb more than just the targeted organism.

For example, neem is one of the most commonly used pesticides among organic gardeners, but studies prove it to be as toxic as some of the synthetic pesticides used for commercial growing. "All natural" does not equate to "safe." Spraying a compound, organic or not, with the intent to kill an organism will absolutely result in harm to other organisms such as insects, mammals, aquatic species (through runoff), or microbes.

IMPROVE MOISTURE RETENTION IN SOIL

I described the importance of organic matter as a source of plant nutrition in Chapter 1, but its benefits extend far beyond nutrition. Soil biota feasting upon organic matter exude organic

TRAP CROPS

>>> A trap crop attracts pests, reducing damage to your food crops. Flowers like amaranth (shown here) can be used to distract flea beetles from food crops (see page 61). Nasturtiums can "trap" aphids. I remove the damaged leaves from my amaranth and use the flowers in bouquets.

compounds that glue soil particles together in clumps known as aggregates. These aggregates are great at holding nutrients but are also full of pores and channels that transport and trap water and air.

When it rains, water is transported through the pores and channels to deeper layers for better long-term storage. Healthy soils rich in organic matter have structures that readily hold enough water to feed plants (water retention) while

Healthy soil clumps together in a way that holds nutrients and just the right amount of water.

GRASS OR MULCH WALKWAYS?

>>> Some gardeners prefer grassy walkways that are maintained with a mower or string trimmer. This design works well with raised beds that offer a physical barrier to keep grass from creeping into the beds, but it's not my preference when creating in-ground beds. Grass from walkways is guaranteed to creep into in-ground beds and will therefore have to be periodically edged out using an edger, digging spade, or shovel.

Mulched walkways must be mulched every other year, but there's little weeding to do. The exception is where the edge of the mulch meets the lawn. You'll have to edge this once per year, but the total linear footage is far less than the amount you'd need if you were edging each individual bed.

Grass walkways are especially striking in the early spring as you await the garden's growth, but keeping the grass out of your beds can be a constant challenge.

simultaneously releasing enough water to prevent soggy roots (drainage). The addition of organic matter, combined with leaving soils undisturbed, is therefore the universal antidote for both dry and overly wet soils.

Stick with no-till methods. Tilling and aggressive digging will pulverize soil aggregates and the beneficial network of pores. Because of their poor structure, tilled soils tend to wash away in heavy rain. We need those aggregates, pores, and organic glues to hold our soil in place, appropriately distribute water, and retain moisture for periods without rainfall.

Don't add sand to improve drainage. Sand will certainly improve drainage in the near term, but it falls short when it comes to nutrient retention. Sand is nutrient poor and does very little for long-term soil health. In contrast, organic matter introduces living and nonliving components to your soil that support proper drainage over time.

Leave roots in place. Remove spent plants by cutting them off at the soil line rather than pulling them out. Leaving the roots to decompose in place deposits organic matter deeper in the soil profile. Soil biota feast upon the dead roots, creating new aggregates and pore spaces, and improving water retention and drainage deep within the soil. (Don't cut off weeds at the soil line, however; they might regrow.)

USE SHEET MULCHING FOR LARGE EXPANSIONS

》》》 Sheet mulching can be scaled to any size area and used to create mulched, easy-to-maintain walkways and nutritious beds all at once. Expanding your garden in this way requires a lot of materials, including cardboard, compost, and, ideally, a large pile of wood chips, wood shavings, or mulch.

Sourcing materials in bulk becomes increasingly important as a garden gets larger, so start learning ways to acquire these products locally for as little cost as possible. Contact local arborists and sawmills about wood chips and wood shavings. Mulch, compost, and/or manure purchased by the yard can often be delivered for a fee, but you may find it cheaper to rent a truck or trailer for a day or two and pick up the material yourself.

Expanding a garden using sheet mulching is no small endeavor, but I cannot overemphasize how well your investment in labor and materials will pay you back in beautifully rich soil. Invest the time now, and you'll be rewarded for years to come.

In planning a large expansion next to my greenhouse, I first smothered the grass with thick layers of cardboard and wood chips, as shown here. I then outlined my bed placement and built sheet-mulched beds directly on top of the wood chips.

Hügelkultur Bed Construction

Select a suitable site for your bed using the site-selection criteria on page 17. Your first hügelkultur bed can be any size, but I recommend starting with an 8 × 3-foot bed that's 2 feet tall at its apex. The materials listed here are approximate—this isn't like baking a cake! It's simply important to keep the relative proportions in mind, especially if you choose to scale up or down.

1 Remove the grass from the area where the bed will be constructed by using a digging spade or a flathead shovel; reserve the grass for building the inner layers. Removing the grass creates a depression in the ground that helps collect water. Alternatively, as with sheet mulching, you can simply smother the grass with several layers of cardboard and build on that.

2 Arrange the largest logs close together in parallel lines on top of the bare ground or cardboard to form the base of the hügelkultur bed.

3 Pile a thin layer of organic matter over the logs to fill any gaps. You want each log layer to fit tightly together.

4 Arrange the smaller logs parallel to the base logs in a way that minimizes crevices. Fill any gaps with organic matter.

5 Pile on a layer of the sticks and twigs. Scatter the blood meal or other source of nitrogen over the sticks and cover them with 6 to 8 inches of mushroom compost or soil. Use your hands or the back of a shovel to lightly pack the soil or compost in place.

6 Cover the bed with a thin layer of straw mulch and plant it immediately, using starts. The plants will keep the soil in place while the bed settles. You can direct sow the following year once the soil has settled.

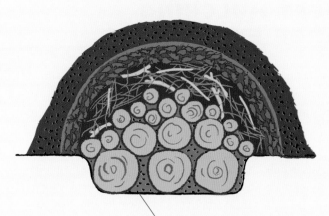

DEPRESSION IN SOIL

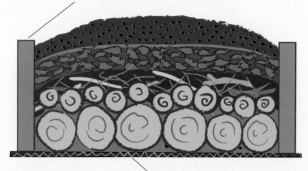

RAISED BED BORDER (OPTIONAL)

CARDBOARD

MATERIALS

- 25–35 feet of 8-inch-diameter logs of various lengths
- 1 cubic yard of organic matter (soil, compost, wood chips, leaves, grass clippings, or wood shavings)
- 35–45 feet of 4- to 5-inch-diameter logs of various lengths
- Large pile of sticks and twigs cut into 2- to 3-foot lengths
- ½ pound of blood meal or other nitrogen source
- 1 cubic yard of mushroom compost or good-quality soil
- Bale of straw or other form of mulch

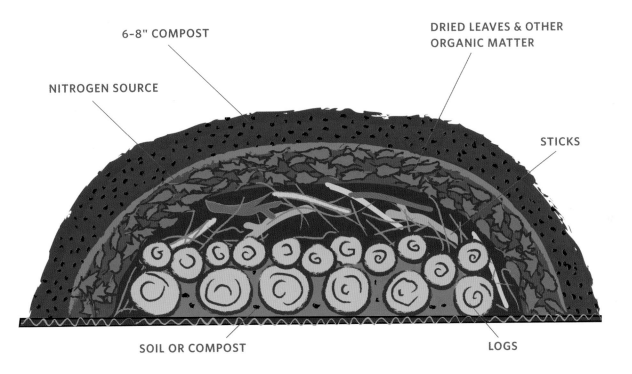

6-8" COMPOST

DRIED LEAVES & OTHER
ORGANIC MATTER

NITROGEN SOURCE

STICKS

SOIL OR COMPOST

LOGS

BUILD A HÜGELKULTUR BED

The term *hügelkultur* ("mounded culture") describes a bed that feels a bit like sheet mulching on steroids. The base of a hügelkultur is lined with logs, smaller logs are piled on next, sticks on top of that, crevices between logs are packed with dried leaves and other organic matter, and a final layer of soil or compost is placed on top. As the logs decay, they form a sponge-like substrate that's capable of storing water while slowly releasing nutrients. Because of this, hügelkultur beds are especially useful in arid regions.

Final bed dimensions are highly variable and depend upon the size of the wood that forms the base. Logs that are just 6 to 8 inches in diameter will create an 18-inch-tall hügelkultur while logs that are between 1 and 2 feet in diameter create a bed height of 3 to 4 feet. As with sheet mulching, there's no set formula. Most any wood will do (including rotten wood; not pressure-treated or painted wood) but I avoid the use of cedar, locust, and other rot-resistant woods because these are so slow to decompose.

ADD NEW CROPS

You now have space to add some easy-to-grow carbohydrate-rich vegetables—beets, turnips, carrots, and potatoes—while experimenting with a few perennial herbs and flowers and perhaps some new annual flowers. I also recommend adding peas and small melons at this stage (see page 63).

A FEW EASY ROOT CROPS

Potatoes, beets, turnips, rutabagas, and carrots have compact growth habits compared to

some of the other carb-rich vegetables such as winter squash, sweet potatoes, and corn. These larger plants are best saved for when you have more space (see Step 7). Beets, turnips, rutabagas, and carrots should be direct sown, while potatoes are grown vegetatively from seed potatoes. A single seed potato is cut into pieces, placed in the ground, and each piece will then sprout a new plant from each eye (see Planting Potatoes, page 66).

These root vegetables are cool-season plants and are therefore planted in the early spring after the soil has thawed and temperatures are warm enough to encourage seed germination. Except for potatoes, these plants can be sown again in the late summer for a fall harvest.

PLANTING ROOT VEGETABLES

Beets are the only root crop that can be started in flats and transplanted into the garden, but you can also direct sow them. Turnips and carrots can be started only by direct sowing. Turnip and beet seeds will sprout in about 5 to 10 days when planted as indicated on your planting chart. Carrots can take up to 3 weeks to germinate. (See Seed Planting, page 34.)

Establishing late summer and fall root crops can be challenging given higher temperatures

Beet seeds were sown between rows of lettuce. As the lettuce is harvested and the beet seeds thinned, beet roots will have plenty of room to mature. See Maximize Production with Succession Planting on page 77 for more information.

and intense sun exposure that cause soils to dry out quicker than in the early spring. Plant around anticipated rain events or plan to water daily when direct sowing in the fall.

When first learning to grow root crops, plant your seeds a bit more densely than the recommended spacing on the seed packet to ensure adequate germination and survival. Thin seedlings when they're about 2 inches tall and have their first or second set of true leaves. Proper thinning

Beet greens are as delicious as beet roots and can be prepared much like chard or spinach.

ATTACK OF THE FLEA BEETLE

>>> Flea beetles are black, flealike insects that jump out of sight the moment you inspect the plant. However, they leave a distinct leaf damage pattern that is a telltale sign of their presence. Infestations are rarely fatal except in the case of young seedlings recently emerged from the ground. Starting plants like tomatoes, eggplant, and pak choi in flats can reduce flea beetle pressure by giving seedlings a place to establish themselves away from the soil where flea beetles reside. Plant the seedlings once they are large enough (at least 10 inches tall for tomatoes and eggplant, and 4 inches for greens) to withstand an attack.

Flea beetles also enjoy the leaves of a few root crops, including turnips and radishes. Because root crops are more difficult to start in flats, I recommend direct sowing root vegetables slightly denser than what is recommended on the seed packet if you find that the emerging seedlings are quickly being devoured. Don't thin seedlings until you're certain that they're surviving.

Flea beetle infestations are easy to identify by the distinctive leaf damage they cause.

Soil Aeration

Aeration ensures that plant roots and soil organisms have plenty of air and space to grow. Aerating no-dig beds isn't always necessary because the construction method naturally encourages the development of pores in the soil.

Aerating, however, seems to be particularly useful for aged sheet-mulched beds built directly on top of compacted clay soils. (Aerating younger beds may be difficult due to the presence of large pieces of organic matter.) Although the mulch's organic matter gets incorporated deeper into the soil over time, aerating accelerates this process.

Aeration is not applicable for hügelkultur beds because the organic substrate used to build the bed will take years to decompose. Hügelkultur beds contain more than enough air and pore space due to the nature of their construction, and they can therefore be left largely undisturbed.

1 Top-dress a vacant bed with compost.

2 Starting at the front of the bed, drive a digging spade deep into the soil and tilt the handle back at a 45-degree angle, just enough to crack the surface of the soil. Some of the top-dressed compost will fall into the crevices and deliver organic matter deeper into the soil.

3 Repeat this process throughout the bed every 10 to 12 inches.

You may be tempted to draw the handle of the digging spade all the way to the ground to watch those enormous aggregates deep within the soil make their way to the surface, where you can easily break them apart. As tempting as it is to manipulate your soils in this way, resist the urge: Let nature do the work for you over time.

of root crops is extremely important because roots need room to expand; they'll never develop correctly if not provided with adequate space.

ADD PEAS AND MELONS

You now have enough space to add in two new annual foods: peas and melons. Both are relatively easy to grow if you get your timing and variety selection right. Peas are a vining, cool-season crop and are therefore planted along a trellis in the early spring as soon as the ground can be worked. Plant again in the late summer for a fall harvest. I've found that spring-planted peas never perform particularly well for me given the position of my garden relative to the sun.

Peas are one of my kids' favorite garden foods.

A BROADFORK: THE ULTIMATE NO-TILL TOOL

>>> You'll someday find that the garden has become too large to aerate using a digging spade. A pronged tool called a broadfork can help. Stand on the center of the broadfork to drive it into the ground and use the long handles to aerate the soil as you would with a digging spade. Broadforks can also be useful when harvesting large numbers of roots or tubers such as carrots, potatoes, or sweet potatoes. Insert the tines of the broadfork far enough away that you don't accidentally pierce the harvest.

My spring garden is always delayed compared to other growers in my region who have better sun exposure. My fall peas produce abundantly. Reference your planting chart to figure out when you should plant peas but don't be afraid to deviate if you find that a more optimal window exists.

Melons have a growth habit similar to that of cucumbers and can therefore be trellised or grown along the ground. Choose small varieties similar in size to cantaloupes (or smaller) that mature quicker than bigger varieties such as watermelons. Kajari melons are one of my favorite varieties for their compact size and the speed at which they mature.

Large melons are more difficult to grow and require a long growing season. Harvest your melons when they smell sweet or the stem leading to the melon has started to turn brown. Don't wait too long or the fruits will crack.

Native flowering perennials like bee balm attract pollinators and serve as important food sources for them.

ANNUALS, BIENNIALS, AND PERENNIALS DEFINED

>>> **Annuals** complete their life cycle within a year. Most familiar garden vegetables are annuals: kale, collards, tomatoes, peppers, squash, zucchini, and others. Annuals die shortly after setting seed. Sometimes, seeds will fall to the ground and germinate as volunteers the following year.

Biennials complete their life cycle in two years. Beets, Brussels sprouts, carrots, and parsley produce foliage during their first year of life and set seed the following year. Distinguishing a biennial from a perennial or annual isn't always easy. Biennials don't necessarily follow an exact timeline: Weather patterns and growing conditions can coax a biennial into flowering before it enters its second growing season, or it can persist for multiple years if grown under highly favorable conditions. Garden biennials are often treated as annuals in that they're most often harvested during their first year of life.

Perennials are planted once and grow back year after year. The majority of perennials die back during the winter, producing new leaves and blossoms the following year. Perennials can persist for decades (and often spread, sometimes aggressively) as long as they're provided with suitable habitat.

GROW PERENNIAL HERBS AND FLOWERS

Perennial herbs and flowers are some of my favorite garden plants because they require very little attention once well established. For potted perennials, prepare the bed as you would for annuals, follow the planting instructions on page 69, and mulch around the newly planted starts to prevent water loss and weeds (see Mulching 101, page 71). To do this, elevate the plantings just slightly above the soil line and fill in the gaps with mulch. Leaving space for mulch keeps it off the stems and prevents rot and disease. Don't forget to water new plantings regularly until they're well established.

Verify that a perennial will grow in your zone. Different perennials tolerate temperature extremes (winter lows and summer highs) to varying degrees. For example, chives are hardy in USDA Zones 3 through 10, and oregano is hardy in Zones 5 through 12. A grower in Zone 4 can reliably grow chives, but oregano is less likely to survive winter. A year with a prolonged or especially harsh winter can push your region into a lower zone, inadvertently causing the death of sensitive perennials.

To avoid this, select perennials whose hardiness zone is at least one level lower than your own zone (for instance, if you live in Zone 7, choose perennials suited for Zone 6 or lower) to ensure survival. Refer to Appendix B to learn more about a plant's hardiness range, but be aware that you can occasionally find cultivars that are more or less hardy than what's listed. Read a plant's label for its exact range. Purchasing perennials from local growers and asking for suggestions guarantees you're investing in plants that will thrive for years to come.

Avoid aggressive spreaders. Certain perennials, such as some members of the Mint family, spread via underground runners that can stretch 2 to 4 feet from the parent plant. Other plants such as catnip (also in the Mint family) aggressively spread seeds. Such plants can require a lot of maintenance to control their spread, and they can be difficult to remove. Perform a quick internet search to verify that you're not planting a species that's going to overrun the garden or that is listed as invasive for your region. Although I offer advice on growing some of the more aggressive perennials in Step 4, I recommend that you start with some of the tamer options discussed in this chapter.

Place your perennials appropriately. Many perennials prefer six or more hours of sun per day; others tolerate full shade. Herbs are best planted as close to your house as possible so you

Elevate your newly planted perennials 1–2 inches above the ground to account for mulch. Piling mulch on the stems can cause rot.

Planting Potatoes

Potatoes prefer light soils. Select smaller potato varieties such as 'Dark Red Norland', small fingerling varieties, or 'Yukon Gold' if you have heavy soils associated with high levels of clay or consistent periods of heavy rainfall. Larger potatoes, like russets, are less likely to reach their mature size in heavy soils.

Plant potatoes in a new location every year to thwart pests like potato beetles that overwinter in the soil. These pests emerge in spring and can easily find a crop that has been planted in the same spot as the previous year. If you find voles and other underground rodents to be a nuisance, grow potatoes in 7- to 15-gallon pots, or install raised beds lined with hardware cloth and plant your potatoes there.

Potatoes are tubers (underground stems) that are traditionally planted in deep furrows, but I prefer a no-dig method that better preserves soil structure and function.

1 Slice seed potatoes into chunks about the size of an egg with at least one eye per chunk. Withered or sprouting potatoes are perfectly suitable. Extremely small potatoes can be planted whole. Spread the potato chunks on a sheet pan or other flat surface and allow them to dry for a few days before planting. Sealing the cut surface helps protect from disease, but it's not critical if you forget to plan ahead.

2 Use a garden hoe or shovel to create a shallow trench about an inch deep and 4 to 5 inches wide. Place your potato chunks in your trench with the cut side down and eyes facing up. Different varieties have different space requirements, but rows are generally spaced 1½ to 2 feet apart with 8 to 12 inches between plants.

3 Cover the potatoes with 2 to 3 inches of compost or manure followed by the dirt you removed from the trench. Mulch with a thick layer of no-spray straw or aged wood chips.

4 Varieties mature at different times, but most can be harvested within 90 to 120 days (smaller potatoes are generally ready sooner; check the recommendation for your particular variety). They are ready to harvest when the foliage starts to die back. Before harvesting, use pruning shears to cut off the tops of the potatoes about 6 inches above the ground and discard the greenery in your compost pile. This simple step makes for efficient harvesting.

5 Hand-dig your potatoes by working your fingers into the soil surrounding the plants, feeling for the outermost potatoes and pulling the entire plant. Collect the potatoes, shake off any excess soil from the roots, and discard the roots in your compost heap. Potatoes can be harvested with a digging spade, but use caution so as to not pierce the tubers.

6 Cure your potatoes before storage. Spread your potatoes out and allow them to dry for a few days out of the sun. I rinse my potatoes before storage, but some people claim that they'll store longer if left dirty. Place your cured potatoes in a cardboard box or other breathable container in a cool (not refrigerated), dry, dark location like a garage. Potatoes can store for up to six months—variety dependent—under ideal conditions.

This 2 × 2-foot wooden frame covered in ½-inch hardware cloth is used as a mobile washing screen for root crops. Rinse heavily soiled veggies and let them dry before storage.

can quickly harvest what you need in the kitchen. Flowering perennials are easier to maintain when planted separately from annual vegetable beds (exceptions to this are discussed in Step 4).

Transplant perennials in early spring or late fall to prevent significant heat and water stress when the roots are disturbed. Use a shovel or digging spade to dig around the circumference of the established plant's rootball and pry the plant from the ground. You'll destroy some of the roots in the process, so strive to leave as much soil intact around the roots as possible. Use a shovel to divide the rootball, if desired, before placing the new perennial in an appropriately sized hole and packing soil in and around its roots.

Transplanted perennials usually appear wilted for about a week until their feeder roots become established. Water the transplants daily, preferably in the morning, or until they stop wilting when the sun is shining. See Dividing Perennials on page 134 for more detailed instructions on digging and dividing perennials.

Harvest perennial herbs. Perennial herbs are harvested slightly differently than annuals in that you don't always harvest the entire plant. Overenthusiastic harvesting can sometimes result in inadequate growth the following year. For example, oregano and chives can be cut down to the ground for their aromatic leaves, because new stems and foliage will grow from the roots the following season. Thyme, sage, and rosemary, on the other hand, should be harvested less aggressively as the following year's growth originates from the stems, not from the roots.

WALKING ONIONS

>>> Walking onions (also called Egyptian walking onions or tree onions) are a fun and prolific perennial to add to your garden. Starts can be ordered online if you can't find them locally. Walking onions produce a flower head of bulbils, a cluster of miniature onions that fall to the ground, take root, and grow into new plants. Dense clusters of bulbils should be thinned or spread out to encourage proper development.

Walking onion greens are tolerant of frost, can persist into the late fall and early winter in milder climates, and can be harvested on an as-needed basis. The bulbs never grow to more than an inch in diameter, making them the perfect substitute for green onions or scallions. You can harvest the entire onion—bulb and all—or just the greens. Leave at least 15 onions in the ground, more for large spaces, to ensure adequate reproduction the following year.

Walking onions produce bulbils, a false flower head filled with tiny onion sets. The bulbils fall to the ground as they mature, giving rise to new plants.

ADD ANNUAL FLOWERS

Annuals can often be coaxed into flowering for nearly an entire growing season and are therefore great additions to enhance diversity, attract beneficial organisms, and beautify your space. The flowers I recommend adding are simple to grow and provide loads of color, heights, and textures in the beds. Many of these plants come in varying colors and sizes, so be sure to select a variety appropriate for your intended use.

The suggested flowers are relatively easy to start from seed and can either be direct sown in the garden or started in flats indoors. Planting seeds indoors gives the plants a jump-start, which is highly desirable for gardeners with short growing seasons or for those who want to ensure flower production earlier in the season. Refer to page 73 for seed-starting instructions.

Space plants appropriately. The exact spacing of annual flowers depends on the mature size of the selected plant. Read the seed packet or plant label for spacing recommendations.

Water plants frequently at first. Planting and caring for annual flowers is much like caring for annual vegetables: They require frequent watering during establishment and prefer moist, well-drained soils.

Deadhead to encourage continuous blooms. Periodically removing spent flowers along with their flower stalks will coax a plant into producing more flowers. Allowing too many flowers to mature into seed heads will signal to a plant that its life cycle is complete, and the plant will die.

Stake top-heavy plants. Wind and heavy rain can wreak havoc on a flower bed if tall plants are not properly supported. Depending on the exact varieties' growth habit, zinnias,

NASTURTIUM

ZINNIA

COSMOS

Sweet alyssum will flower late into fall even after experiencing multiple frosts.

Calendula's attractive blooms can be dried, infused into oils, and used to make healing skincare products such as salves and lip balm.

sunflowers, tithonia, and cosmos often benefit from being staked.

Enjoy cut flowers indoors. Frequently harvesting flowers encourages continuous blooms. Zinnias, sunflowers, snapdragons, tithonia, calendula, cosmos, and marigolds, among many others, make great cut flowers.

Annual flowers can reseed. Zinnias, sunflowers, calendula, and tithonia reliably reseed when dried flower heads are left in the garden. These volunteers may lose certain traits like size or color due to cross-pollination between varieties, but this isn't necessarily a bad thing, just something to note.

a few easy annual flowers

- Calendula*
- Chrysanthemums*
- Cosmos
- Dianthus*
- Nasturtiums
- Pansies/ johnny jump ups*
- Petunias*
- Snapdragons*
- Sweet alyssum*
- Tithonia (Mexican sunflowers)
- Zinnias

*Some annual flowers are considered cool-season plants. Calendula, for example, is a cold-hardy annual that I plant in early spring and again in fall. Fall plantings produce flowers well into December in my region.

MULCHING 101

Mulch reduces water loss, suppresses weeds, protects soil from too-hot sun and extreme temperatures, introduces valuable organic matter, and maintains a tidier garden. On the other hand, mulches can attract slugs, insects, snakes, and voles. The benefits usually outweigh the negatives, so observe what's happening in your garden and adjust your mulching strategy as needed.

Soil microbes feast on mulch, temporarily pulling nutrients, especially nitrogen, out of the soil while doing so. Although these nutrients are released into the soil as the mulch decomposes, the temporary depletion means slow-to-decompose mulches like wood chips are less than ideal for annual beds. Annuals have a relatively short life cycle and therefore a short window to acquire their nutrition. Mulch annuals with quick-to-decompose substrates like pesticide-free straw, compost, or aged wood chips.

Perennials are less susceptible to temporary variations in nutrient availability because they persist year after year; nutrient acquisition happens over the long term rather than just a single season. The best choice for mulch around perennials and walkways is therefore one that is slower to decompose (hence doesn't have to be replenished as often) and is preferably abundant and free.

A FEW MULCH OPTIONS

Agricultural waste products like pecan hulls or coconut fibers are regionally available products that are often abundant and slow to decompose. Ask about the farm's growing practices and its use of pesticides before applying these products to your garden.

Fresh wood chips from arborists and commercial mulches are slow to decompose and are therefore best used in walkways, borders, and perennial beds.

Aged wood chips, sawdust, and wood shavings are quicker to decompose than fresh wood chips. Allow wood chips to age for one to two years before applying to annual beds.

Straw and hay can introduce weed seeds into the garden. To prevent this, leave the bales outside and keep them well watered to encourage the seeds to sprout. After a few weeks, break apart the bales, destroying the seedlings in the process, and use the hay or straw as intended.

Pine straw and dried leaves are regionally available for little to no cost at different times of the year. Pine straw can be used around perennials but it can be difficult to weed, as the weeds become tangled in the needles. Dried leaves work best as a mulch when shredded, but they can be applied whole if shredding isn't an

Before renewing a layer of mulch, use a digging spade, shovel, or edger to remove any grass that's crept into the sides of your mulched borders.

option. Sprinkling a bit of straw or some wood chips on top of the leaves will prevent them from blowing away.

Bulk mushroom compost (or any type of compost) is my favorite type of mulch for vegetable beds, especially in areas where I plan to direct sow rather than put in starts. The seeds will germinate directly in the compost as long as they're given enough water. Chunkier mulches like aged wood chips and straw have to be cleared off the planting surface before you direct sow.

MULCHING DOS AND DON'TS

Never pile mulch directly on top of weeds or grass. First remove any weeds or smother the vegetation with cardboard. Freshly weeded garden beds require just a couple of inches of mulch, established walkways require 2 to 6 inches, and areas newly smothered in cardboard require 8 to 12 inches.

Refresh mulch every other year. Healthy soil builds up underneath mulch as it decomposes. This becomes the perfect place for weed seeds to take hold if you don't add another layer of mulch as needed.

Mulch under and around fences. Weeds can easily grow up through the fencing, where they become nearly impossible to remove by mowing, with string trimming, or by hand. These weeds not only look messy but can also become a source of weed seeds in the garden. Slide cardboard under the fence (or better yet, have the cardboard in place before installing the fencing) and bury the bottom few inches of the fence along with the cardboard. The mulched border should extend from the fence at least 18 inches.

Edge mulched borders once per year. Grass and other vegetation will creep into your mulched edges, and you'll need to use an edger, digging spade, or shovel to reclaim the mulched border annually. Placing bricks, logs, rocks, or other physical barriers along the perimeter of mulched areas can help prevent this.

Heavily mulched beds can attract burrowing rodents like voles that damage roots. Protect your crops from burrowing rodents by planting in raised beds that have ¼-inch hardware cloth across the bottom.

START SOME SEEDS

Growing starts from seed saves money and allows you to select from a wider range of varieties, so it's a skill worth mastering despite the challenges. Even those of us who have been starting plants from seed for years have difficulties. Start experimenting now when you can affordably replace any casualties, and within a few years, you'll be experienced enough to fill your entire garden with plants that you've grown from seed.

CHOOSE EASY PLANTS

Extremely small seeds like celery can be difficult to grow in flats because the tiniest of seeds give rise to the tiniest of seedlings. Smaller seedlings are generally more difficult to care for. Other seeds, like tomato and pepper, require warm soil temperatures to germinate. I therefore recommend two lists of seeds for beginning seed starting. The first includes larger-seeded annual flowers and vegetables that are some of the easiest to grow indoors. You'll have the greatest chance of success with these plants. The second list features plants with small seeds that germinate easily but require more attention to reach maturity.

easiest to grow

- Cucumber
- Summer squash
- Zucchini

a bit more challenging

- Beets
- Chard
- Collards
- Kale
- Lettuce
- Peppers
- Tomatoes

USE THE RIGHT POTS

Use small to medium pots filled with a good seed-starting mix. Standard four- or six-pack pots are best for beets, chard, lettuces, kale, collards, tithonia, nasturtiums, sunflowers, and calendula. Single pots that are 2 to 4 inches in diameter are best for summer squash, zucchini, and cucumbers. These plants grow quickly and need the extra growing space. Don't choose pots smaller than recommended: The smaller the pot, the more difficult it is to control moisture levels.

Gently used pots can often be gathered for free at garden centers, especially big-box stores with garden centers. Give the pots a quick wash, then let them dry and disinfect in the sun before use. Alternatively, buy reusable plastic or silicone pots or trays from garden supply stores. Reusable plastic trays are much thicker than single-use trays and should last for about a decade if well

Beets can be sown in flats or directly in the ground in clumps of three. Multi-sow beets, radishes, leeks, and even onions to save space.

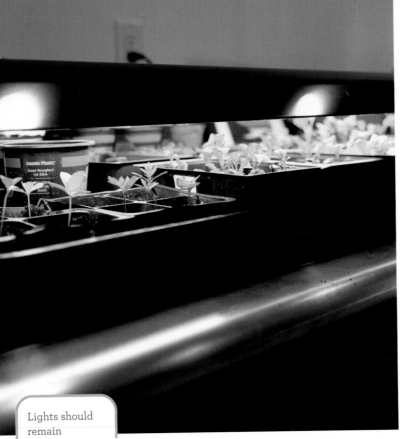

Lights should remain suspended just a couple of inches above your starts. Move them up as your plants grow.

(temperature, type of soil, specific plant, and so on), so it's best to keep an eye out for mild plant wilt and soil dryness. You can also compare the weight of your flats before watering and again after. Flat weight is one of the best indicators of water needs. Plants often prefer to dry out a bit between waterings rather than be given water unnecessarily.

All seedlings need light. The easiest way to grow indoors is to use LED grow lights or basic fluorescent lights strung just a couple of inches above your seedlings. A grow light timer precisely provides your plants with at least 14 hours of sun each day. If you don't have lights, place your plants directly in front of your sunniest window.

Move them outdoors as often as the weather allows (after they've germinated and when it's sunny and temperatures are above 40°F [4°C]) to help them get even more light. Plants that are not provided with artificial light or transported outdoors often (even when grown in a very sunny window) will become leggy and weak.

Some seeds need heat. Tomato, pepper, tomatillo, and several other seeds won't sprout unless soil temperatures remain at a steady 70° to 80°F (21° to 27°C). These temperatures are difficult to achieve in the winter and early spring when starting seeds indoors. Seed-starting mats can be placed underneath flats to provide low, constant heat and to encourage germination. Cool-season plants like lettuces, kale, and collards will germinate without heat, but germination is improved with it.

Seed-starting mats outfitted with thermostats are best because you have the most control over temperature. Ones without thermostats work but they have the potential to get hotter than desirable. Always keep an eye on soil moisture levels when using heat mats as the soil will dry out more quickly.

cared for. Start with 50-cell trays and move on to 72-cell trays once you're having success. Last, you can always use items from the recycling bin such as single-serving yogurt cups, empty egg cartons, or even empty toilet paper rolls. Be sure to poke holes in the bottom of vessels to allow for drainage.

Use a high-quality seed-starting mix rather than potting soil. Seed-starting mixes are lighter and are usually better at retaining water compared to a basic potting soil. Do not use soil directly from the garden, as it's often too dense.

TAKING CARE OF SEEDLINGS

Water them carefully. Seedlings are sensitive to both drought and saturation. The goal is to ensure that the soil is moist every morning before the plants are subjected to light. The frequency of watering depends on a number of factors

Starting Seeds in Pots

Refer to your planting chart or seed packets to determine when to plant in your area. People are often tempted to start seeds indoors too early, but this results in plants that are too large for their pots and unhealthy. Likewise, starting seeds too late in the season won't allow enough time for the plants to reach maturity.

1. Fill each pot to the rim with soil, but don't pack it down. The soil will naturally settle after being watered, and it's important to maintain a lip above the soil line that can collect water. Gently water the pots before seeding if the soil is particularly dry. Label the pots.

2. Use your finger or a pencil to make a small divot at a depth appropriate for that particular seed. Drop in one or two seeds, then brush a bit of the surrounding soil over the seeds or, for shallowly planted seeds, sprinkle fresh soil on top.

3. Water the flats using a very gentle spray or place them in trays of water for bottom watering. Some people prefer the latter method, as it can prevent wilt diseases as the seedlings develop. Seedlings are sensitive to saturation, so be sure not to oversoak them. It's better to let the flats dry out a bit between waterings.

4. The frequency of watering depends on several factors (temperature, type of soil, specific plant, and so on), so watch for mild plant wilt and soil dryness. Check in the mornings to be sure the soil is moist before the plants are subjected to light.

Harden them off before planting.
Plants that are grown indoors must not be abruptly planted outdoors. The sun can burn the leaves and even kill the plants in some cases. Start by moving your plants into the sun for a few hours each day and incrementally increase this time by a few hours over the course of a week. Plants that have been grown without lights and instead have been moved in and out in accordance with the weather won't likely require a hardening-off period because this will have been accomplished over the course of their development.

Plant at the right time. Plant your seedlings outdoors once temperatures are appropriate, seedlings have been hardened off, and the plants have good root development. Gently slide your plants out of their pots to see if the roots and soil slide out together and retain the shape of the pot.

Roots should be abundant and well-developed, but not tightly wound. Return the plant to its pot if the roots are underdeveloped and the soil falls away from the roots. Wait a few days before checking for improved root development. If your plants are root bound, lightly massage the roots free from binding before placing the rootballs in the ground.

Check to be sure seedlings have well-developed rootballs before planting them in the garden.

SOURCING SEEDS

>>> There's no shortage of options when it comes to buying seeds. I prefer to support smaller organic and/or biodynamic seed companies, preferably a local one. Some of my favorites are High Mowing Organic Seeds in Vermont, Sow True Seed in North Carolina, Turtle Tree Seed in New York, Wild Garden Seed in Oregon, Uprising Seeds in Washington, Fruition Seeds in New York, MIgardener in Michigan, Backyard Garden Seeds in Mississippi, and Southern Exposure Seed Exchange in Virginia.

A company located in a climate like yours often offers an abundance of varieties well suited for your region. (Purchasing from such businesses is a suggestion and by no means a requirement.) Likewise, I look to some of the northeastern companies for cold-tolerant varieties because I know their winters are much harsher than my own.

Seed catalogs and websites are packed with choices. Some of the more interesting and visually enticing varieties like frilly kales adorned with unique color patterns are tempting, but I recommend reading the description before purchasing. As a beginner gardener, you're looking for words and phrases such as *reliable, easy to grow, hardy, prolific, tolerant,* and *resistant*. I certainly choose a few interesting varieties from time to time, but I often come back to tried-and-true, reliable producers.

Three stages of successionally planted bush beans can be found in my summer garden: new sprouts like these, beans that are getting ready to produce, and plants that are nearly done producing.

MAXIMIZE PRODUCTION WITH SUCCESSION PLANTING

Plants rarely produce for an entire growing season. Bush beans, for example, produce beans for four to six weeks, then the plants die. You can achieve a steadier supply of beans throughout summer by using a technique known as succession (or staggered) planting, in which you sow seeds of the same crop on different dates. Using the bush bean example, you would sow beans once every four to six weeks to produce a regular supply for several months.

Succession planting also ensures that you're not harvesting tremendous amounts of any one vegetable at any given time. Many gardeners focus on planting long rows of a single crop and then preserve whatever they're unable to consume. While this is certainly a great technique to fill your pantry with put-up food for winter, it's not as practical as spreading out the plantings and eating your vegetables fresh. I still aim to have excess for preservation (when applicable), but a staggered harvest means there's less work involved, as there's a smaller amount of a crop to process at any one time.

Succession planting allows you to extend the harvest window for many vegetables.

WHAT AND WHEN TO PLANT

The number of succession plantings you're able to grow each season depends on the total number of frost-free days and the number of days it takes a crop to mature. For example, Zone 3 averages 120 frost-free days per year while Zone 7 averages 170. More frost-free days means a greater number of plantings each season. Quicker-to-mature crops like lettuces, radishes, and kale can be sown as often as every three weeks starting in early spring and stopping in late fall.

Slower-to-mature plants such as bush beans, zucchini, summer squash, sunflowers, cucumbers, beets, carrots, turnips, and potatoes can

be planted every four to six weeks. And because there's always a rolling turnover in the garden, some of these plants—zucchini, summer squash, beets, and cucumbers—are best started in pots. This way you're not using valuable garden space until that crop is ready to be planted out. Replace one spent crop with a new and different crop each time.

easy crops for succession planting

- Beets
- Bush beans
- Carrots
- Cucumbers
- Kale
- Lettuce
- Potatoes
- Radishes
- Turnips
- Zucchini

Crops like tomatoes, peppers, and okra aren't often planted in succession because these plants need an entire growing season to mature. Succession planting these crops requires a growing season with very few days of frost.

A STEP FURTHER: INTERPLANTING

Succession planting can be taken a step further by anticipating a plant's replacement early on. For example, I know that my early-spring spinach will bolt when temperatures begin to rise, which means I need a replacement on standby. I can either start that replacement in a pot or plant something like bush bean seeds between my spinach rows. The spinach will be clipped off at the roots by the time the bush beans take over. Planting in this way requires some knowledge of space requirements and longevity, but it's an excellent technique for maximizing production in a small growing space.

Early-spring lettuces were strategically harvested to make space for a developing cauliflower.

SEASONAL GARDEN CLEANUP

Fall cleanup. As the season ends and frost claims the last of the summer bounty, you'll want to tuck away the garden for winter. Stakes and trellises are best stored out of the weather to prolong their life. Diseased plants should be removed and isolated from the compost pile and garden to prevent spreading infection to next season's plants.

Winter and early-spring cleanup. Leave nondiseased plants in place in winter to help protect the soil. Start the cleanup a few months before it's time to plant to guarantee that the beds will be prepped and ready. Clip dead plants off at the soil line and add the plant residues to the compost pile. Plants that leave less residue, like lettuce, can be left in place to decompose then smothered with compost in spring. Most perennials die back during winter, and the dead debris should be removed before the perennials begin to sprout again in spring.

FRUIT TREES
and shrubs

Expanding your garden to incorporate perennial fruits offers endless possibilities. I've limited my discussion to common perennial fruits that grow in a wide range of climates, but many other fruits may be suitable for your region. Some of the more interesting selections may help you overcome challenges such as weather patterns, topography, soil conditions, and herbivore pressures. There's much success to be had if you're willing to combine a bit of experimentation with patience.

Plum blossoms are some of the first flowers to appear in early spring.

the plan

- **SELECT A SITE WITH PERMANENCE IN MIND.** Perennial fruits can tolerate nutrient-poor soil, but they'll fail to thrive in waterlogged or especially dry soils. Select a site with good drainage, access to water, adequate sun, and plenty of space to accommodate mature plants.

- **EVALUATE YOUR PROGRESS.** As you strategize the placement of perennial fruits, think about whether you're satisfied with your current efforts or whether you need to make adjustments. By now you have a better understanding of the factors that impact gardening success. You've invested a large amount of time (and likely money) in your current garden, but if something isn't working, don't continue to put effort into it.

- **CHOOSE FIVE PLANTS.** Select three different fruit trees, such as apple, pear, and plum (two of each); four blueberry bushes (at least two different varieties); and a single blackberry or raspberry variety (three to five plants, enough for a 10- to 20-foot row).

- **TALK WITH LOCAL GROWERS** to learn what performs well in your region, then choose varieties suitable for your hardiness zone.

BEFORE YOU DIG IN

Trees and shrubs take time to establish themselves. Think long term when starting a mini orchard. Don't expect fruit production for at least two years as the plants establish healthy roots. Devote time to proper planning, too: Most fruit plants are difficult to relocate once they reach a certain size. Blueberries and other shrubs can be moved, but this is less true for trees or spreading perennials like raspberries.

Don't fret about soil quality. Fruit trees and shrubs prefer moderately nutritious soils. They're better at scavenging for nutrients than shallow-rooted plants because their roots reach greater lengths and depths. Amending the soil around tree roots is nearly impossible; the plants must learn to thrive on the native soil. Top-dressing with compost or an organic fertilizer around the root zone in spring can help improve plant health, but there's no reason to prepare a tree site in the same way that you'd prepare a vegetable garden bed.

Give trees and shrubs lots of sun. Fruit production is energy intensive, which means fruit trees and shrubs need to harness as much of the sun's energy as possible. Blueberries, blackberries, raspberries, currants, gooseberries, and pawpaws can produce fruits with as little as six hours of sun per day, but yield will improve with increased sun exposure. Most other fruit trees and shrubs prefer eight or more hours of sun daily.

Fix drainage problems before planting. Very few fruit trees, shrubs, and vines can tolerate excessively moist soils. Refer to Test Soil Drainage, page 18, before planting to ensure that the soil drains quickly and efficiently. To encourage water to move away from roots rather than pool around them, you can build up soil to alter the land gradient and thus redirect the flow of water before planting the trees or shrubs.

Ensure access to water. In spite of the warning above, fruit trees will fail to survive in dry soils, especially during that first year of growth, so you'll need access to a hose or other water source. See page 91 for a new technique, swales on contour, in which water is collected in and around the garden, trees, or shrubs to ensure the ground absorbs as much rainwater as possible.

LOOK FOR FROST POCKETS

Cold air moves downhill and settles in low points or areas where air movement is obstructed by trees or buildings. Frost pockets—small, often

Most fruit trees, including this peach, produce far more blossoms (and fruits!) when placed in a sunny location.

isolated areas where cold air becomes trapped—can exist in valleys, on hillsides, or in urban settings. These pockets can experience frost even though the surrounding area is frost-free. This becomes problematic for gardening when daytime highs coax fruit trees into producing blossoms too early. If the blooms experience frost damage overnight, the tree may not produce fruit, or the fruits will be poorly developed.

Understanding how topography impacts air movement is important in deciding where to plant fruits and other sensitive perennials. Ideally, you will avoid frost pockets, but this isn't always possible depending on the size and position of your property. If frost pockets are a risk, look for fruit varieties that bloom later in spring. For example, an early apple like 'Gala' blooms long before a late-season apple such as 'Pink Lady'. Choosing late-season fruits may mitigate the effects of a damaging frost.

DIVERSIFY PLANTINGS FOR PRODUCTION

In the previous chapter, we explored the benefits of diversity for building a protective food web (flowers attract insects, insects attract birds,

and so on), but now let's look at the benefits from a production standpoint. When designing an orchard, no matter how big or small, focus on planting different types of trees and shrubs, especially in the beginning. Planting just a few of each type of fruit will help you figure out what works and what doesn't before you invest a large amount of money. Then plant more of the ones that perform well.

For example, I replaced two poorly grown apricots that never produced fruits. I've since learned that even though apricots are hardy in my zone, they rarely produce fruits. Initially investing in more than just two apricots would have been an even costlier mistake.

The other benefit to diversifying your orchard is that it can balance fruit ebbs and flows. It's common to experience varying levels of success with different crops from year to year. Your pears may experience a low-production year while your apples throw out a bumper crop. Pruning, which I'll discuss later, can help even out annual production, but some years there's little you can do to control success. Planting a diversity of trees and shrubs helps to ensure greater cumulative success over time.

FROST POCKET

Topography or large obstructions like trees and buildings can trap cold air, creating frost pockets.

CHOOSE THE RIGHT VARIETIES

Some trees and shrubs fail to produce fruits even when your hardiness zone seems appropriate for that particular plant. For example, some fig varieties, such as 'Chicago Hardy', are hardy down to Zone 5. The plant returns year after year as a perennial does, but it will only produce fruits in Zones 8 and above. 'Brown Turkey' figs more reliably produce fruits in colder regions. Ensure that you're selecting a plant that's not only hardy but that also reliably bears fruit in your region.

Visit a local farmers' market to ask farmers what has worked for them and use their expertise to inform your own efforts. There are thousands of varieties of apples, yet most of us are familiar only with a few of the more common, such as 'Red Delicious', 'Gala', or 'Granny Smith'. Local growers can tell you about lesser-known varieties that are better suited for specific needs, such as 'Pink Lady' (a winter storage apple), 'Arkansas Black' (a colorful sauce apple), or 'GoldRush' (a flavorful green apple resistant to mildew and fire blight).

RESEARCH CHILL HOURS

"Chill hours" describes the number of hours that a tree must remain dormant at 35° to 45°F (1.5° to 7°C) or lower in order to flower in spring. In general, figs, citrus, and quince require a low number of chill hours compared to apples, peaches, and plums. Plants with lower chill-hour requirements will flourish only in warmer climates. Plants that require a higher number of chill hours need a colder climate to successfully bear fruit.

Some fruit plant varieties can be grown only within a particular climate, while others have been bred to expand their normal range. For example, a 'Honeycrisp' apple requires 800 chill hours; an 'Ein Shemer' apple requires just 350. These two varieties represent the upper and lower limits of chill-hour requirements for apples, which allow for apple production over diverse regions. So in addition to considering a plant's hardiness zone and ability to reliably produce fruit, make sure to select a variety whose chill-hour requirements are suitable for your region.

Don't cling to a mistake just because you spent a lot of time making it. Replace, remove, or relocate trees or shrubs that don't appear to be thriving.

KNOW POLLINATION REQUIREMENTS

Cross-pollination (the transfer of pollen from one plant to another) is required for certain trees and shrubs to produce fruits. These types of trees and shrubs must be planted with a pollenizer. Without one, they'll produce very little to no fruit at all. In some cases, the pollenizer can be the exact same variety, such as a 'Granny Smith' pollinating another 'Granny Smith'. But most cross-pollinators prefer to be planted with a different variety (for instance, a 'Gala' apple with a 'Honeycrisp' apple).

Every now and again, you'll come across a variety that accepts pollen from a pollenizer, but its own pollen is useless because it's sterile. The sharing of pollen is therefore unreciprocated. For

Pears are insect pollinated and must be planted near another pear variety to produce fruits.

FRUIT VARIETY SELECTION CHECKLIST

>>> Different fruit varieties are bred for traits such as size, disease resistance, early fruit production, late fruit production, long storage life, and/or thornless vines (as is the case with blackberries and raspberries). There are benefits to each trait, but you'll need to know what you're buying to plan accordingly. As you research varieties that best suit your needs, ask yourself:

- Is this variety hardy for my region, and do I have enough chill hours for it to produce fruits?
- Do I have two cross-pollinating varieties that produce flowers at the same time of year?
- Are my varieties compatible pollenizers?
- How large will my tree or shrub grow, and do I have adequate space?
- Have I selected a self-supporting blackberry or raspberry, or do I need a trellis?

There are no hard-and-fast rules that apply to all fruits, so knowing the specifics of each chosen variety is extremely important.

instance, if you were to plant a 'Jonagold' (sterile) with a 'Gala' (pollenizer), the 'Jonagold' would bear fruit but the 'Gala' would not. In this case, you'd need a third variety such as a 'Cortland' to ensure fruit production in all three varieties. When purchasing trees that need a pollenizer, select two varieties (three if one of the varieties is sterile) that flower at the same time of year and are known to be compatible pollenizers. To ensure proper pollination, plant them no more than 50 feet apart.

Trees and shrubs like peach, fig, and some cherry varieties are considered self-pollinators, meaning they don't require a pollenizer. Although fruit yield is often improved in the presence of a second variety, having that second plant is not essential. Listed here are some common cross-pollinators and self-pollinators.

cross-pollinators

- Apple
- Blueberry
- Pawpaw
- Pear
- Plum
- Sweet cherry

self-pollinators

- Apricot
- Blackberry
- Citrus (most)
- Fig
- Mulberry
- Nectarine
- Peach
- Raspberry
- Serviceberry
- Sour cherry

SELECT PLANTS THAT FIT THE SPACE

Trees and shrubs can take up a lot of space at maturity. Pruning helps control the size and shape of a tree or shrub, but forcing a standard-size tree (about 20 feet tall, variety dependent) to stay shorter than 10 feet tall isn't recommended. The stunted aboveground

'Bonanza' peach is a dwarf variety that grows to just 4 feet tall. The fruits have less flesh compared to other dwarf varieties, but its compact nature makes it an interesting addition to the home garden.

biomass will be inadequate to support a full-size root system, and energy will be pulled away from fruit production as the tree ages. If you want small trees, select dwarf and semi-dwarf varieties.

A columnar apple tree grows in a 2- to 3-foot-wide column and stops growing at just 8 to 10 feet tall. Columnar trees can be tucked into small spaces or used as interesting accent plants.

GENERAL SPACING RECOMMENDATIONS FOR SELECT TREES AND SHRUBS

The following size and spacing recommendations are intended as general guidance to help you get started. Some plants will grow to be taller or shorter than what is listed. For example, most standard peaches grow to just 15 feet tall. Refer to each plant's information label or tag for more exact size and spacing information.

PLANT TYPE	HEIGHT AT MATURITY (IN FEET)	SPACING RECOMMENDATION (IN FEET)
Standard tree	18–25	30–35
Semi-dwarf tree	12–15	15
Dwarf tree	6–10	10
Columnar apple	8	3–4
Fig	8–12	15
Blackberry, trellised	10	5
Blackberry, self-supporting	3–5	3–5
Raspberry, trellised	5–6	3–4
Raspberry, self-supporting	5–6	3–6
Blueberry, rabbiteye	15	15
Blueberry, highbush	6–10	8
Blueberry, half-high	2–4	2–5
Blueberry, lowbush	1 foot	2

Positioning plants too close together increases the likelihood of disease and reduces fruit production. And when competing for limited space, enthusiastic spreaders like raspberries spread more aggressively by shooting runners farther from the parent plant in an attempt to acquire necessary nutrition.

At planting time, space the plants according to their mature size. Using trees and shrubs as hedges or borders, training plants to grow along a fence or the side of a building (see Espalier Fruit Trees, below), or selecting columnar apples are techniques for successfully fitting more varieties into a smaller space.

ESPALIER FRUIT TREES

>>> Espaliered trees are pruned and trellised to grow along a single plane, which is especially useful for gardeners working with limited space. Espaliered trees can be trained to grow along a south-facing fence or building wall or used to create a border or fence. Espaliered trees are available for purchase, or you can espalier your own. Young bare-root trees are best because you can train their growth early on. Older trees are less pliable and more difficult to train. Dwarf varieties of pears and apples are most frequently used, but it's possible to use smaller trees like figs and some cherry varieties.

Choose the desired final shape and build a corresponding support structure. You can use a variety of structures, including wooden posts with wire strung between them or an existing structure such as a fence. Certain branches are pruned away and the remaining branches are tied to the support, where the wood hardens in place as it matures. Espalier pruning videos can be found online for more information.

Espalier trees can be trained to grow in many shapes, but candelabra, horizontal cordon, and fan are three of the most common.

USE SWALES TO CAPTURE RAINWATER

Swales on contour implement a unique water-holding strategy. Start by digging a ditch—a swale—that's capped at both ends and can therefore collect water during rainstorms. Then pile the dirt from the ditch on the downhill side of the swale to create a berm that serves as an elevated garden bed or as a barrier to keep rainwater from moving downhill. Swales follow the contour of the land, and when properly designed, rainwater evenly fills a swale and is slowly absorbed into the surrounding soil. Trees and shrubs placed downhill from the swale benefit from increased water availability.

Constructing swales can be a labor-intensive process but ultimately saves time when it comes to watering and irrigation. Swales work well for growers living in arid regions or for those who are growing on a hillside or mildly sloped area that drains quickly and efficiently. When deciding whether to build swales, consider how your topography, climate, and geology impact water storage and drainage on your property.

Keep in mind that collecting water in and around a building, septic system, or drain field is undesirable. Swales should therefore be placed at least 15 feet away from these structures. Low-lying areas, yards with poor drainage, extremely steep slopes, or locations with particularly high water tables should also be avoided. I don't build swales on the creek side of my property where drainage is more of a concern than water retention.

Swales can either be filled with porous organic matter such as wood chips or straw or left open and planted with a perennial cover crop such as clover, vetch, or rye. Swales filled with organic matter can be used as a walkway and serve as a source of slow-release nutrition as the mulch slowly decomposes.

The swale will have to be cleaned out every five or so years for it to continue collecting water. Leaving a swale open is a lower-maintenance option, but the swale needs to be constructed at a gentler slope to prevent the swale from caving in on itself.

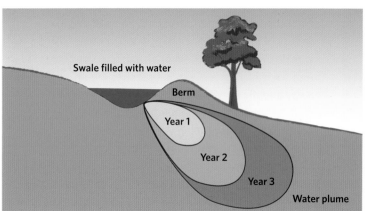
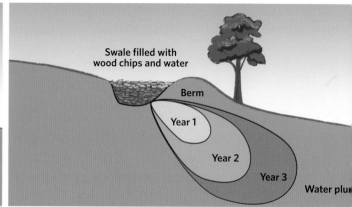

The walls of a swale left open should be constructed at a gentler slope than one filled with organic matter. Water availability will increase over time as the water plume reaches greater depths and lengths.

Swale on Contour

Place a swale on a sloped area (mildly sloped is fine) where plenty of rainwater flows across the soil surface. Swales must follow level contour lines because water will always move toward the lowest point; water will pool at one end of a swale if it's not perfectly level. Many online videos demonstrate how to identify and mark contour lines.

The depth and width of swales are highly variable, ranging from 6 inches deep and 18 inches wide to 2 feet deep and 5 feet wide. Larger swales are typically machine built. Unless you have access to digging machinery, I suggest starting with smaller swales that can be constructed using hand tools. See how the technique works for your location, then expand upon the construction once you're certain that swales on contour are useful in your situation.

1 Mark the contour lines using flags or other markers. The length and spacing of the contour lines should account for proper spacing of the new fruit trees and shrubs to be planted.

2 Dig a ditch 10 inches deep and 20 inches wide along the contour line, piling the soil on the downhill side to create a berm. If you plan to use the berm as a garden bed, place cardboard beneath the pile of soil to kill the existing vegetation.

3 When building a small swale, determine if it's level by filling it with water from a hose (this is impractical when constructing larger swales). The water should fill the swale evenly rather than pool at one end. Alternatively, and even better, observe the swale during a large rain event, then adjust the swale depth to ensure water doesn't pool to one side.

4 Fill the ditch with wood chips or mulch to serve as a walking path and source of nutrition for the soil.

5 Smooth the berm into an even mound if you plan to use it for garden beds and top-dress the soil with compost. Plant trees on the downhill side of the swale, at least 6 feet from its base.

John Henry Nelson of Stone and Spade Earthworks installed fruit trees, shrubs, and flowers along a swale that serves as a passive watering system.

AN INTRODUCTION TO BERRIES

Berries of all kinds are a fun and prolific addition to the home garden. For the sake of simplicity, I've narrowed my discussion to a few of the most common berries: raspberries, blackberries, and blueberries (strawberries are covered in Step 4). I recommend starting with these, then branching out to lesser-known varieties like serviceberries, currants, honeyberries, and goji berries as you're ready. As with all fruit trees and shrubs, berries often take a few years to start producing abundantly, so be patient in planning future expansions.

Blackberries have been one of my most successful fruit crops and are enjoyed by the entire family.

Berries have chill-hour requirements ranging from 200 to 1,000 hours. Most berries fall within the 400- to 600-hour range, but some, such as 'Dorman Red' raspberries or 'Brightwell' blueberries, require just 200 to 300 hours. For this reason, blueberries, raspberries, and blackberries aren't typically grown in south Florida and other regions close to the equator.

Mammalian and avian pests have potential to be among your biggest challenges. Pressures from birds, squirrels, raccoons, deer, and other foraging animals may call for erecting protective infrastructure—such as tall fencing to exclude deer or bird netting draped over ripening fruits. Some growers go so far as to build large cages lined with chicken wire to surround their entire berry patch.

HOW TO GROW RASPBERRIES AND BLACKBERRIES

Raspberries and blackberries are available as self-supporting or trellised, with or without thorns. These plants must be pruned every winter, without exception, so be sure you're willing to do the work before building too large a bed. Start by creating a 10- to 15-foot row of each (no more than 20 to 30 feet in total) and expand as you're ready. Rows should never be shorter than 10 feet long due to plant-spacing requirements. Space rows 8 to 10 feet apart. Blackberries and raspberries can be easily propagated from existing plants if you eventually want to expand.

Some varieties are prolific spreaders. Red and yellow raspberries and, to a lesser extent, some blackberries aggressively spread via underground runners. New plants can pop out of the ground up to 10 feet from the parent plant. This is a great feature if you're interested in encouraging your fruits to spread, but it can become laborious when trying to control growth. For this reason, spreading raspberries and blackberries are best planted in an isolated area with a mowed perimeter. Don't place them within 10 feet of other crops or perennial beds.

Other varieties don't spread as much. Most blackberries and some purple and black raspberries produce far fewer runners. Instead, their canes bend to touch the ground and produce roots where the cane tip meets the soil. Controlling their spread is fairly easy, as you can see exactly where the next plant will grow. To prevent spread, keep the cane tips off the ground or clip off any roots that appear. Dispose of the roots in a fire or by leaving them in the sun to dry for at least a month so that they don't accidentally take root in a compost pile or a deserted area of your yard.

Fruiting times differ. Summer-bearing raspberries produce new canes in spring each year. Blackberries produce new canes in midsummer. Both go dormant in winter and yield fruits in summer on the previous year's canes, then those canes die back. This means you won't get berries the first year; you'll have to wait until the following year when those canes bear fruit. Remove the second-year canes once they have finished fruiting.

Ever-bearing raspberries also produce canes in spring, but the tops of the canes produce berries that same year in late summer. So yes, there's potential for fruit that first year, but year-one flowers are best removed to push energy into the developing root system. Remove the upper portion of the cane when it's dormant and expect a smaller yield from the lower part of the canes the following summer. Ever-bearing raspberries produce more berries but the plants are harder to maintain because you have to prune the same canes twice—once to remove the top portion

and again when the cane is spent. (See Pruning and Trellising Blackberries and Raspberries, page 105.)

Growers often plant a small patch of ever-bearing and a small patch of summer-bearing raspberries to extend their season of production. Plant the two in separate areas. Raspberries are so prone to spreading that the two types can become intermingled over time. Knowing which canes to prune and which to leave behind becomes confusing once the two weave together.

Flowering blackberries put on quite a show in early summer. Follow spacing and pruning recommendations to encourage abundant yield.

STEP-BY-STEP

Constructing Trellises

Blackberry and raspberry trellises don't need to support loads of weight; trellises simply keep the plants' canes upright and prevent them from falling over. Because the trellis posts will be exposed to moisture in the ground, they should be made from cedar, locust, or another rot-resistant wood; untreated pine will rot quickly. You can use any type of wood for the arms.

The following materials will build a 10- to 20-foot-long trellis. You'll need to install additional posts for bed lengths greater than 20 feet. Rent or borrow the post-hole digger if you don't anticipate needing this tool often.

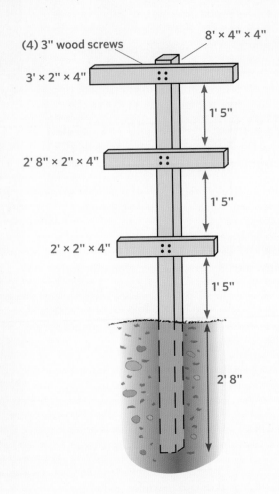

(4) 3" wood screws

8' × 4" × 4"

3' × 2" × 4"

1' 5"

2' 8" × 2" × 4"

1' 5"

2' × 2" × 4"

1' 5"

2' 8"

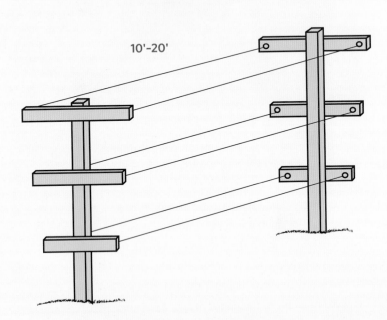

10'-20'

String sturdy wire between the eyeholes to complete the trellis.

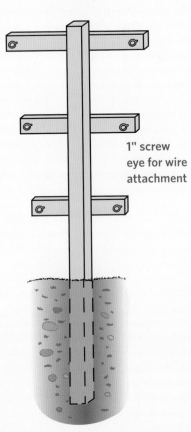

1" screw eye for wire attachment

MATERIALS & TOOLS

- Post-hole digger
- 2 (8-foot) rot-resistant 4×4 posts
- Level
- Circular saw
- 2 (8-foot) lengths 2×4 lumber
- Drill
- 24 (3-inch) exterior wood screws
- 6 (1-inch) screw eyes
- Screwdriver
- Wire cutters
- 60–120 feet 9- to 12-gauge galvanized wire
- Leather gloves

1 Use the post-hole digger to make two holes 32 inches deep and 10 to 20 feet apart, depending on desired row length and number of plantings. Rows longer than 20 feet require additional posts.

2 Place a post in one hole, straighten it with a level, and repack the soil using a stick or other object to tamp the soil into the hole. Adding some rocks or gravel to the hole can improve sturdiness. Repeat with the second post.

3 Using the circular saw, cut each 2×4 into three pieces: one 3' 2", another 2' 8" inches long, and a third 2' 2" (discard the remaining 6-inch scrap). Drill pilot holes for the wood screws in each arm. Attach three arms to each post using 3-inch wood screws (spacing and screw placement is indicated in the diagram).

4 Drill pilot holes for the screw eyes as indicated. Twist each eye into place using a screwdriver pushed through the eyehole.

5 Cut the wire to appropriate lengths (2 feet longer than the distance between posts) and string the wire between the eyes by looping it through the eyehole and twisting it onto itself. Wearing leather gloves, pull the metal wire as tightly as you can before attaching it to the opposite screw eye. Trim off the excess wire.

A BIT ABOUT BLUEBERRIES

Choose from four types of bushes, each with varying mature heights—rabbiteye, highbush, half-high, or lowbush. Rabbiteye bushes grow to 15 feet tall and 10 feet wide, producing as much as 15 pounds of berries from a single bush. Highbush blueberries generally mature at 5 to 8 feet tall, with some varieties growing as tall as 12 feet. Half-high bushes mature at 2 to 5 feet, and lowbush varieties are just 6 to 24 inches tall.

Choose varieties suitable for the allotted space and note their bloom time. Blueberries are self-pollinating, but the yield is greatly improved when two, preferably three, different varieties are planted together. Your selected varieties should have overlapping bloom times if they're to exchange pollen.

As with many fruits, blueberries form on last year's growth. Plants therefore produce minimal yield for the first two or three years as the plant develops. Remove the flowers the first year after planting to encourage energy to move to the roots.

Blueberries are more sensitive to soil nutrition than other berries. They prefer acidic soils, which can be encouraged by annually mulching with 3 to 4 inches of pine bark or wood chips. Mulching blueberries annually also helps to suppress fungal and bacterial diseases as mulch smothers spores and keeps the soil cool (high temperatures encourage soilborne pathogens). When planting blueberries or other acid-loving plants, place them in the ground as you would any other tree or shrub. Allow them to establish themselves and watch for signs of stunted growth or yellowing leaves that indicate improper pH (confirm your suspicion with a pH test).

Plants are thought to exude compounds through their roots into the soil to help adjust the pH around the root zone, which means that your blueberries will likely thrive without attempts to adjust soil pH. Besides, most soils are naturally acidic. However, if pH appears to become a problem, fertilize with a sulfur-based fertilizer and top-dress around the plants with animal manure. Do this annually until you see an improvement. Search your state's Cooperative Extension website to diagnose specific issues with blueberries and/or other fruits and vegetables.

DON'T USE PEAT MOSS

>>> Peat moss is a common ingredient in seed-starting mixes and is frequently used as a soil amendment to acidify soils for plants like blueberries that prefer a low pH. Peat is harvested from peat bogs, which are estimated to contain as much as one-third of the world's soil carbon. Because peat is generated at a rate of just 1/16 inch per year, centuries-old peat is being harvested and consumed by gardeners and agriculturists at a rate far beyond which it can be replenished. Harvesting methods put peat bogs at risk of peat fires, which can release tremendous amounts of carbon into the atmosphere. I therefore avoid the use of peat for any application in the garden. Look for potting soils that contain coconut coir rather than peat and use sulfur-based fertilizers to help adjust soil pH.

PURCHASE HEALTHY PLANTS

Trees and shrubs can be purchased from local nurseries, at big-box and grocery stores, or from online nurseries that ship plants. Plants from local nurseries are often the most expensive, but they are among the healthiest. If these plants are out of your price range, order plants online or buy them from chain stores immediately after the stores receive shipments during gardening season. The longer plants stay at the store, the greater the chances of neglect.

Many fruit trees are grafted onto a rootstock and are impossible to grow from seed. If a tree *can* be grown from seed—for example, a paw-paw—you may have to wait 10 to 15 years to achieve fruit production. Trees (and shrubs and vines, for that matter) are more typically purchased as plants in one of three forms.

Container plants are grown in pots and have well-established root systems. These plants are often older, which means they'll produce more quickly compared to the following options. Container plants are the most expensive, but I highly recommend buying container trees if you can afford it. They're easier to plant, have higher survival when cared for properly, and produce fruit in a fewer number of years.

Bare-root plants are dug from the ground when they're dormant (not actively growing). The roots are rinsed, and the plants are refrigerated without soil. Bare-root plants look like sticks with a few small roots on the bottom. Some growers feel that bare-root plants are best because they establish themselves in the soil where they'll spend their entire life.

Even if this is true, bare-root plants are extremely young, meaning that trees like apples and pears can take five or more years to bear fruit. Bare-root vines and shrubs such as blackberries and blueberries need only two or three years to start producing and are therefore a better candidate for purchasing as bare root, especially as they're so much cheaper than container plants.

Balled-and-burlapped plants are grown in the ground before being removed (usually with a machine) and having their rootballs, complete with soil, wrapped in burlap and wire. Balled-and-burlapped trees and shrubs come in larger sizes, but their roots have been heavily disturbed during the digging process, and thus new plantings are more sensitive and need more TLC when getting established.

If possible, avoid balled-and-burlapped trees and shrubs: They are similar in cost to a container tree but are more difficult to establish. One reason is that keeping a balled-and-burlapped tree or shrub well watered in a nursery is challenging. The water beads off the burlap, especially when the rootball becomes dry. Balled-and-burlapped trees and shrubs may be suffering from dehydration without showing signs of it.

> *Plants from local nurseries are often the most expensive, but they are among the healthiest.*

Planting Fruits

Trees, shrubs, and vines are best planted in fall or early spring when the plants are dormant and temperatures remain low. Container trees and shrubs can be planted anytime, but you'll need to give them extra water. Bare-root or balled-and-burlapped plants should never be planted in summer, and most nurseries that ship bare-root plants send them within a planting window. With very few (or compromised) roots to support their growth, bare-root plants are more likely to become stressed and die.

1. Bare-root plants must be revived before planting. Soak the roots in water for two to six hours before placing them in the ground.

2. For container and balled-and-burlapped plants, dig a hole three to four times as wide as the rootball and as deep as the roots. The sides of the hole should gently slope upward to catch water and to allow space for root development. A hole for bare-root plants should be twice as wide as the rootball with a depth equal to the length of the roots and have the same saucer-shaped edges as for container plants.

   The root collar—the transition line between trunk and roots—on the tree, shrub, or vine should never be planted belowground. Soil or mulch sitting on the root collar can cause rot and disease. Trees, shrubs, and vines are better off being planted too shallow than too deep. Elevate the root collar a few inches above the soil line if you plan to mulch around the base of the tree or shrub.

3. Set the plant in the hole. Holding the plant upright, start filling the hole with soil that you dug out. Pack soil directly around the roots to ensure proper positioning. Once the plant is straight and even, lightly pack the soil back into place. Use compost to fill the final inch of the hole.

4. Mulching around the base of trees and shrubs is optional except with blackberries and raspberries, which should always be mulched as there's no way to mow around canes; excessive weeds will choke out your fruits.

5. Water the plants well and keep them moist. Newly established trees, shrubs, and vines need a lot of water their first year as their roots grow into the surrounding soil. Unlike the starts that you place in vegetable beds, trees and shrubs don't exhibit signs of water stress until that stress has been prolonged. If you don't get frequent rain, water the rootball and the 3 feet of ground that surrounds the trunk or stem once per week for the first year of the plant's life.

   Note: Some plants require more attention than the broad recommendations outlined here. Read each plant's information label or tag and follow any special planting instructions.

Mulching around trees and shrubs is recommended because it helps prevent weeds and simplifies yard maintenance. Mulching is not a requirement, but your decision to mulch must be taken into account when determining how deep to plant the rootball.

PLANTING A CONTAINERIZED TREE

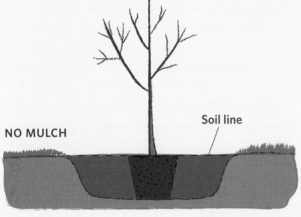

NO MULCH

Soil line

Fill to the top of the rootball if not mulching

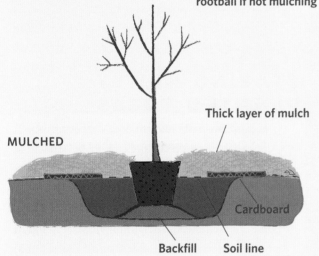

MULCHED

Thick layer of mulch

Cardboard

Backfill   Soil line

DON'T USE FERTILIZERS OR SOIL AMENDMENTS.

Trees, shrubs, and vines can thrive off what's already available in the soil by harnessing nutrients with roots that span great lengths and, sometimes, depths. Heavily amending or fertilizing a tree's hole before planting can burn the roots and cause plant death. It encourages the roots to circle around and grow in the rich, amended soil rather than extend into the native soil where nutrient acquisition may be more challenging. This ultimately results in poor root development and nutrient depletion around the rootball.

PLANTING A BARE-ROOT TREE

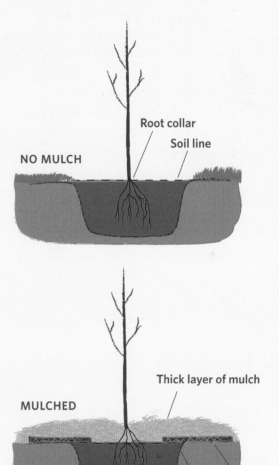

NO MULCH

Root collar

Soil line

MULCHED

Thick layer of mulch

Cardboard

Backfill   Fill line

DO TOP-DRESS WITH COMPOST.

Trees, shrubs, and vines benefit from an annual topdressing. Apply 1 or 2 inches of compost or animal manure beginning as early as that first year. Start the topdressing 4 inches from the trunk or stem and spread the compost as far out as the longest branches into the area known as the crown. The breadth of the branches often represents the breadth of the roots, which means the nutrients from the compost will leach into the soil directly on top of the roots.

PREPARE TO PLANT

While you don't need to amend tremendous amounts of soil before you plant fruit trees and shrubs, you should take certain steps before you start digging holes. For actual planting methods, see Planting Fruits, page 98.

Measure twice, dig once. Use a tape measure and flags (or some type of movable marker) to mark the placement of the trees, shrubs, and vines. Leave enough space to accommodate the mature size of each plant and follow spacing recommendations. Avoid planting too close to buildings, power lines, underground infrastructure such as septic drain fields, and areas that may become shaded in coming years due to nearby vegetation.

Build necessary supports. Even self-supporting blackberry and raspberry varieties can benefit from being trellised, especially once the canes are weighted down by ripening fruit. You can certainly try growing a self-supporting variety without using structures, but don't be surprised if you later find that supports are necessary.

STAKING NEWLY PLANTED TREES

>>> Only stake young trees if they need it. Trunks and roots develop best when allowed to develop unassisted—the mild stress from swaying encourages plant tissues to grow stronger. However, larger trees or trees planted in a windy site may benefit from one or two stakes to keep them positioned upright.

Most trees shorter than 8 feet tall require a single 5-foot stake driven into the ground at a 45-degree angle. Cross the stake against the trunk and attach it using wide, stretchy tape or some type of soft, thick cord. Abrasive or thin cords such as twine or rope can damage the bark. A tree's vascular system lies directly beneath the bark; damage can kill the tree. To prevent rubbing, wrap the cord so that the trunk is not in contact with the stake.

Stakes can usually be removed after one growing season once the roots have become well established.

CREATE A TREE GUILD

The area around a tree can be planted with a tree guild—a collection of beneficial perennials that support the tree's health. For example, you might plant comfrey or borage at the drip line—the edge of the tree's canopy where rainwater falls—and place more drought-tolerant plants like garlic chives directly under the tree. A well-designed guild contains plants that attract pollinators and other beneficial insects, deter wildlife, add nutrients and organic matter to the soil, repel insects, and suppress weeds.

These proclaimed benefits of planting a tree guild are not unanimously agreed upon by scientists (or gardeners), but there's no doubt that increasing diversity in the garden is always beneficial. A tree guild may or may not provide a tree with more nutrition, but it will beautify your micro-orchard, create more garden space for perennials, and build a beneficial food web. You can plant anything you'd like in a tree guild, but some common perennial additions are listed here.

Comfrey and perennial alliums such as these nodding onions can be planted beneath a fruit tree to build soil health and to deter pests.

tree guild plants

- Borage
- Chives
- Comfrey
- Daffodils
- Garlic
- Oregano
- White clover

DON'T BE AFRAID OF PRUNING

Learning to prune can feel intimidating, and you may worry that you'll damage a plant beyond its ability to repair itself. Fortunately, this is not the case. My experience is that you're more likely to damage a plant beyond repair by withholding pruning and allowing excessive uncontrolled growth early on. I did this with a set of bare-root pear trees and forever regretted the way in which those early branches developed.

Think of pruning like a haircut. Regularly scheduled haircuts improve hair quality just as regularly scheduled pruning improves plant quality. If you get a haircut that you don't particularly like, you wait a few months for it to grow out and start afresh. The same is true for trees and shrubs. Time, patience, and new growth can almost always cover up any pruning mistakes. Neglect is less easily masked.

PRUNE REGULARLY AT THE RIGHT TIME

Most fruiting trees and shrubs need annual pruning to encourage higher fruit yield, improve air

circulation (thus forestalling disease), prevent a plant from growing too large, and beautify the perennial trees and shrubs. Most common fruits can be pruned based on the following information. If you want to be meticulous, research a particular tree or shrub's needs before pruning.

For example, you'll learn that plums grow aggressively and can be pruned at most any time of year to control growth. But prune cherries in

Pruning your trees, shrubs, and vines should be done annually, without exception, to encourage healthy plants and prolific fruit production.

AVOID USING WEED BARRIER

>>> Applying weed barrier, a.k.a. landscape fabric, is tempting when planting trees in a neat and orderly fashion, but these fabrics impede the development of healthy soil. The mulch that you place on top of landscape fabric will decompose over time, creating a healthy layer of organic matter that's physically separated from the native soil below. Worms and other soil-dwelling organisms that help draw organic matter deeper into the soil profile are unable to pass through the fabric to reach this rich, organic layer, so the tree never benefits from the nutrients in the mulch.

Furthermore, this top layer of organic matter eventually becomes the perfect growing medium for weed seeds. Landscaping fabric will suppress weeds for about a year or two, but then you'll be right back to pulling weeds. You're better off using a thick layer of cardboard covered with a thick layer of mulch around the trees. The short-term result is the same as using landscape fabric, but you'll be better supporting soil health.

Healthy soil becomes trapped on top of landscape fabric, which prevents plant roots from accessing the soil's nutrients.

summer, because they're less susceptible to a bacterial disease in the warmer months. Peaches are best pruned when they're flowering, as the wounds heal quickest at this stage. Use the following guidelines to get started, but fine-tune your techniques as you're interested or able.

Prune most branches when the tree is dormant. Nutrients from the leaves are drawn into the tree trunk when temperatures start to drop in fall. This energy remains stored during winter and moves back into the branches when temperatures rise in spring. Remove woody limbs and branches—those that are older growth and therefore less pliable—when most of the energy is stored rather than when that energy is being reallocated toward growth. That means pruning branches in late winter before the tree or shrub starts to bud in spring.

Remove dead branches any time of year. They are, however, often easiest to identify in summer when the rest of the tree is covered with leaves.

Summer prune to control growth. Cut away young, nonwoody branches in summer so that the tree doesn't allocate resources toward growth that will eventually be removed. Prune only the current year's growth, not growth from previous years.

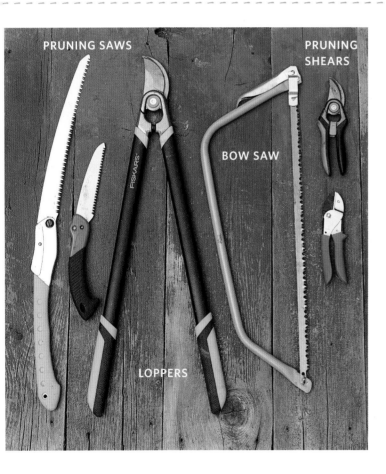

PRUNING SAWS

PRUNING SHEARS

BOW SAW

LOPPERS

PROPER PRUNING TOOLS

›››   To prune correctly, you need the right equipment. Always sterilize tools by wiping the blades with rubbing alcohol before pruning to prevent the spread of disease. Dried sap and dirt can be removed by scrubbing clippers with a soapy scouring pad.

Pruning shears or clippers are used on branches that are thinner than ½ inch in diameter. Bypass clippers (the two blades slide past each other) make cleaner cuts than anvil-style clippers (the blade hits against a flat surface).

Loppers can handle branches from ½ inch to nearly 2 inches in diameter, depending on the length of the lopper handles and the size of the blades. Longer handles generally provide better leverage, so those models can cut larger branches.

Pruning saw or bow saw removes branches that are too large to be cut with loppers. A pruning saw can fit into spaces a bow saw is unable to reach.

WATCH OUT FOR FRUITING SPURS

Most varieties of apples, pears, cherries, and plums bear fruit on short, stalky branches called fruiting spurs. There are varietal exceptions to this, but fruiting spurs are more common than not. Fruiting spurs produce blooms in early spring and will do so for as long as 10 years. Learn how to identify these branches so as not to accidentally remove them while pruning. The only fruiting spurs that should be removed are those that are crowded too closely together.

Fruiting spurs most often appear on second-year wood (first-year wood in rare instances) and in some cases are so short that they're easy to miss. When learning to prune, avoid removing any branches that appear particularly short or stalky, then observe the flowering period of the tree to verify that you've correctly identified its fruiting spurs.

PRUNING BLUEBERRIES

Blueberry pruning is often ignored but done right, pruning will improve the plant's health and ensure heavy fruit production year after year. Prune blueberries in late winter. Start by removing dead, dying, or diseased branches (this can be done any time of year). Next, remove any stems that cross or are growing toward the center of the bush by cutting them off at ground level. Finally, cut out the oldest stems until you've removed one-third of the bush—including the branches you removed in the previous steps.

Older stalks are thick and woody; young stalks are thin and pliable. Each stem will produce berries for about four years, so removing older stems is essential to create space for new, fruit-bearing stems. In the first three years of growth, limit pruning to dead and dying stems, as all young stems produce fruit.

Many trees bear blossoms on fruiting spurs. Get to know these short, stubby branches so you can avoid accidentally removing them while pruning.

PRUNING AND TRELLISING BLACKBERRIES AND RASPBERRIES

>>> Remove spent blackberry and raspberry canes by cutting them off at the base starting as soon as the fruits have been harvested. The canes are most often removed when the plant is dormant.

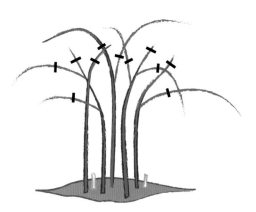

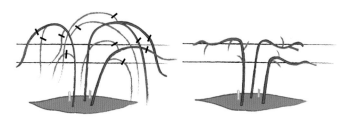

Self-supporting blackberries and nonspreading raspberries. Thin new canes to the strongest five per plant. Trim the tips so the plant is no more than 4 feet tall. The tips can be pinched off in summer to control growth. Trim lateral branches to 12 inches and remove any that appear crowded.

Trailing blackberries. Thin new canes to the strongest four per plant. Trim the tips so the plant is no more than 10 feet long. The tips can be pinched off in summer to control growth. Trim lateral branches to 12 inches and remove any that appear crowded. Wrap the dormant canes around support wires.

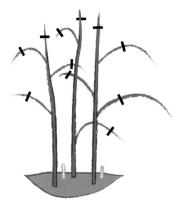

Summer-bearing raspberries. Thin new canes to 6 inches between canes with 12-inch-wide rows. Trim the tips so the plant is no more than 4 feet tall. The tips can be pinched off in summer to control growth. Trim lateral branches to 12 inches and remove any that appear crowded.

Ever-bearing raspberries. Remove the spent tops of ever-bearing canes after fruiting; remove the remainder of the cane the following year once fruiting is complete.

Fruit Tree Pruning

Prune no more than 20 to 30 percent of a tree or shrub at one time. The leaves convert sunlight into food while the roots collect necessary nutrients and water from the soil. The roots are growing at a rate similar to that of the foliage, which means that a heavily pruned canopy produces inadequate energy for the roots. Maintain a balance between these two important functions to ensure good tree health and bountiful fruit production.

1 Determine the tree's natural disposition to grow with or without a central leader. For a central leader, choose the strongest, most obvious vertically growing trunk (or branch) and remove all others that try to grow vertically. For an open canopy, remove any central leaders, especially those appearing in the center of the tree.

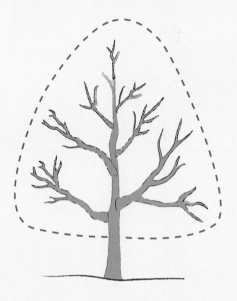

Central leader—Pears and apples prefer a central dominant stem. Even when you remove the central leader, the tree will produce vertical stems to replace the lost lead.

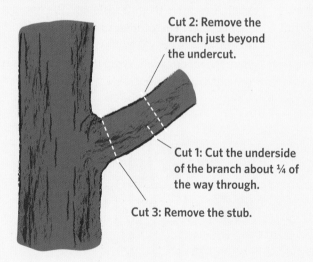

Cut 2: Remove the branch just beyond the undercut.

Cut 1: Cut the underside of the branch about ¼ of the way through.

Cut 3: Remove the stub.

Tip: Cutting off a large branch can result in the bark being pulled downward from the trunk under the weight of the falling branch. To prevent this, remove large branches in a three-step process: First undercut one-quarter of the way through the branch to prevent the bark from tearing. Second, cut through the top of the branch to remove the limb. Finally, remove the stub.

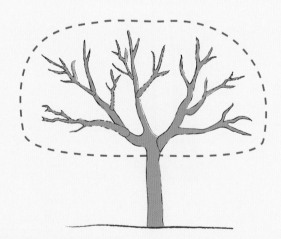

Open canopy—Peaches, plums, and cherries prefer an open canopy in which the central leader is removed and the center of the tree remains unoccupied.

Tip: The joint where two branches meet has a transition zone where bark starts to grow outward from the dominant stem and upward into the new stem. Cut just beyond this collar to encourage the bark from the dominant stem to grow over the stem nub. The bark will eventually encapsulate the nub, which promotes healing and prevents entry of disease at the site of the cut.

Tip: Upward growth should be removed directly above an outward-facing bud to encourage the tree to grow out from the center of the tree or shrub rather than up or in toward the center. Cut at a 45-degree angle just above the bud to prevent water from collecting on the cut tip. Slope the cut down and away from the bud to stimulate bud growth.

2 **Remove branches with any of the following issues.**

Diseased, dying, or dead. There's no reason to allow a tree or shrub to continue pouring energy into a limb that isn't likely to survive because of disease. Remove diseased branches to prevent spread.

Crossing or touching other branches. Crossing branches impede airflow and light. As these branches age, they have the potential to rub against one another, which can damage the bark and the tree's circulatory system directly beneath the bark.

Growing toward the center of the tree. Aim to keep growth moving outward and away from the tree's center. Branches growing inward will impede air circulation and light.

Growing too close together. Branches should be spread out to encourage good air circulation and adequate spacing for fruit development. Strive for branches 6 to 18 inches apart. Try to envision what smaller branches will look like as they mature and remove branches that will eventually create a tightly spaced canopy.

Growing horizontal or near horizontal. The weight of mature fruits can overload a branch and break it. Aim to have the branches no more than a 60-degree angle from the trunk.

Competing as central leaders. Choose the strongest central leader for pear, apple, and other trees that prefer this growth habit. Some branches will try to compete for this position; remove those that do.

Need redirecting. New branches often grow outward from the tree then start to grow upward toward the tips. These branches can be forced to continue growing outward by removing about a third (just a few inches on dwarf and semi-dwarf trees) of the branch directly above a bud that's facing the direction you want the branch to grow. Removing the tissue above this bud will force it to grow outward in a more desirable direction. Most new branches are pruned in this way to better control the shape of the tree.

STEP

4

edible and flowering
PERENNIALS

Expand your garden by 400 square feet to allow space for more edible and flowering perennials. Perennials return year after year, often with minimal effort, making them an important feature in a garden. You may choose to plant the entire expansion with perennials or to leave space throughout your new beds to plug in annuals.

The beds that line my house boast borage, astragalus, milkweed, yarrows, perennial alliums, sage, chives, thyme, lilies, and even a blueberry bush. Vacancies in the bed are filled with annual herbs, vegetables, and flowers every spring.

the plan

- **BUILD YOUR BEDS.** Decide which soil-building method you'll use to create raised or in-ground beds: sheet mulching, hügelkultur beds, swales on contour, or some combination of these techniques. You may want to create this expansion incrementally as you acquire plants—say, 100 square feet in winter, another 200 square feet in spring, and a final addition in fall. Acquiring perennials takes time and/or money, so try not to rush it.

- **EXPLORE LANDSCAPING DESIGN.** Intermingling perennials with annuals gives a garden improved visual appeal. The new beds can be irregularly shaped and fluid to accent your home or existing property features. Plant most of the new beds with perennials and leave open blocks to be filled with annual flowers or vegetables.

- **CHOOSE PLANTS AND PLACEMENT WISELY.** Plant perennials that are edible, are native to your region, have medicinal qualities, attract pollinators, or are aesthetically pleasing.

BEFORE YOU DIG IN

Adding perennials to diversify and beautify your garden is an ongoing effort. However you decide to complete this step—adding the entire expansion at once or slowly adding perennials as you build up the rest of your garden and see suitable placements for new plants—know that this expansion is about growing food, beautifying your space, and improving garden diversity with low-maintenance additions.

Choose plants based on sunlight requirements. Perennials can be grown in the shade, partial sun, or full sun. Sun-loving perennials prefer six or more hours of sun per day, while shade-loving perennials prefer fewer than six. Most sun-loving perennials won't tolerate being grown in a shade garden (or vice versa), but some, like coralbells, can go either way. Unlike annuals, perennials will often accept a wider range of sun variability.

Know the soil moisture. Some perennials will tolerate moisture extremes while others won't. Sedums, cacti, lavender, yarrow, thyme, catnip, daylilies, sage, and coneflowers can tolerate dry conditions. Ligularia, cardinal flower, ferns, elderberry, hardy hibiscus, and some irises can tolerate wet conditions (although some irises despise having wet feet). Choosing a site with moderate moisture can host a wider range of plants.

Select varying colors and bloom times. Flowers attract pollinators. Perennials generally flower for just six to eight weeks at approximately the same time each year. Gardens filled with varying colors and bloom times attract a wider range of pollinators for a longer period. Some flowers, like daffodils and tulips, flower for just four weeks in early spring. Chives and oregano come next. Sunchokes, Joe-Pye weed, and goldenrod (the latter two being native plants in some regions) are among the last to bloom in autumn and keep the late-season pollinators satisfied.

Choose plants that bloom at different times to ensure color and attract pollinators throughout the entire growing season. Look for interesting

Some perennials such as coralbells can tolerate both shade and sun.

Choose perennials with different bloom times to attract pollinators throughout the season. These red poppies are some of the first to bloom in early spring.

foliage, too: Plants like bronze fennel with textured and colorful leaves provide a garden with more depth and color even when the plants aren't blooming.

Allocate space carefully. Don't waste time, energy, and money on plants that you don't particularly enjoy or edibles that you'll never eat. Direct those resources to your favorites, and only plant large swaths of edible perennials when you'll actually use the plant. For example, planting more than one or two rhubarbs may prove to be a waste of space unless you're a rhubarb fanatic with an efficient rhubarb preservation method. That space could instead be allotted to more walking onions, garlic, or asparagus or to other favorites.

Plant in spring or fall. Perennials are best planted when lower temperatures won't stress new plants. Pot-grown perennials can be put in the ground at any time of year as long as you're sure to water them every few days while their roots get established. Perennials that have been dug up are best planted in late fall after they've flowered and when temperatures are low. Fall planting gives the perennial enough time for root establishment before it goes dormant, thus resulting in a vigorous plant the following spring.

When planting bulbs, check the label to determine whether you have a fall- or spring-planted variety. Some bulbs can be planted at any time while others have a seasonal preference.

PERENNIALS THRIVE ON FEW INPUTS

>>> Nutrients harnessed by perennials throughout the growing season are transferred to the roots and/or woody shoots as the plant dies back and loses its leaves at the end of the growing season. Perennials are always recycling energy that was collected in previous years. And because the soil is rarely disturbed around their roots, the soil biota remain active and productive, further aiding in unassisted plant health.

Perennial beds should always be installed using the soil-building techniques recommended in previous chapters—sheet mulching or hügelkultur beds—but there's no need to apply compost every spring as you would with annual beds. Compost additions can be limited to once every three or four years, or when you notice the plants needing an extra boost.

Most perennials need to be divided every few years (see page 134) to keep the plantings controlled and to encourage flowering, but they largely take care of themselves once established. Perennials that aren't thriving are usually planted in an area that's too wet, dry, sunny, or shady. Struggling perennials can certainly experience more complicated issues, but always start by moving a dissatisfied perennial to a new location to see if it's happier and healthier elsewhere.

If you miss the fall window, plant dug perennials in early spring. Transplants should never be planted in the heat of summer. Keep transplanted perennials well watered until their leaves remain upright and turgid in the heat of the day, indicating that the roots have established themselves and are taking up adequate water.

ROOTS IMPROVE SOIL

Plant roots exude compounds into the soil that feed soil microbes, aid in plant nutrient acquisition, and pump in carbon-rich substances that improve the soil's structure and nutrient-holding capacity. As roots die, they become an important food source for microbes and other soil-dwelling decomposers that transform dead organic matter into usable nutrition for plants. Encourage root growth in the soil as often as possible to keep soils alive and productive.

Perennial root systems that persist year after year are especially good at stimulating soil processes. Perennials often go dormant in winter, but their roots are still performing their essential tasks of supporting the soil environment for next year's growth and preventing erosion. Some gardeners prefer to plant perennials and annuals together, as short-lived annuals can benefit from the long-term support of the perennials.

I prefer to plant the bulk of my annual vegetables in rows in my main garden while leaving space to intersperse annual vegetables and flowers throughout my perennial beds. This approach allows for simplified maintenance. However, interplanting perennials and annuals can be especially useful in erosion-prone areas: Perennial roots stabilize the soil, making it usable for growing food.

Plants with deep root systems, such as borage (shown here) and comfrey, harness and transfer minerals from deep within the soil profile. The plant tissues are composted and thought to supply high levels of minerals.

USE COVER CROPS

Cover crops offer some of the same benefits as perennials but without the permanency. For example, replacing a spent annual crop with a new annual crop (see Minimize Production with Succession Planting, page 77) keeps the soil active, but this isn't always practical or possible depending on the season. In these instances, a cover crop can protect soil from the sun's radiation, stimulate soil processes, prevent erosion, suppress weeds, enhance nutrient levels, and generate carbon-rich plant debris to feed the soil.

Sow cover crops by scattering the seeds on top of the soil and raking them lightly into the soil with a garden rake. This is by no means exact, but I aim to have approximately one seed per every 4 square inches of soil. Refer to the specific sowing recommendations for more exact guidance. Keep the seeds watered while they are establishing themselves.

Annual cover crops are easy to control. They can be removed using one of three kill methods: chop and drop, winterkill, or smothering. To chop and drop flowering plants such as buckwheat and annual clover, cut the stems an inch or two above the soil line using a string trimmer when the plants are in full bloom. Let the stems and flowers dry out on the surface of the soil before planting that bed again. The roots feed the soil as they decompose and the dried material acts as mulch.

Cold-hardy crops, such as tillage radishes, oats, and field peas, are planted in late summer and early fall and left unprotected in winter. In many climates, these crops aren't hardy enough to survive an entire winter, so they die off after repeated heavy frosts. This cover crop management technique, called winterkill, is the simplest way to control the growth of cover crops while priming the beds for spring planting. The winterkilled debris is left in place in spring and top-dressed with compost.

Smothering is used for those rare instances in which you've planted an annual cover crop that, for one reason or another, has become difficult to remove (or if you have beds that have become completely overrun with weeds). If hand-weeding is not a viable option, you can kill off the vegetation by smothering it with cardboard or sheets of thick black plastic, sometimes called silage tarps. Ideally, this is done in fall or winter so that you can use that bed the following spring.

If using cardboard, pile compost on top to hold the cardboard in place and plant new crops in about two months once the cardboard starts to decompose (some people pull the cardboard out from under the compost if they're not satisfied with its rate of decomposition).

If using a silage tarp, weight it down with sandbags or rocks and allow it to remain in place for two to three months while it kills virtually everything beneath it. The plants beneath the tarp are deprived of sun, and the soil becomes extremely hot. Some growers place silage tarps over their entire garden all winter long because it

BUCKWHEAT: THE PERFECT COVER CROP

>>> Buckwheat is my favorite cover crop: It's simple to sow, grow, and control, which makes it a great starting point. A quick-growing annual flower that produces a large amount of biomass in just six weeks, buckwheat has attractive white or pink blooms (the white form is cheaper and produces more prolific flowers) that attract pollinators and make excellent cut flowers.

When the blooms start to deteriorate, chop and drop the stems (don't wait until they set seed or they'll overpopulate your garden) to help feed soil microbes and to act as a mulch.

SPECIFIC BENEFITS OF DIFFERENT COVER CROPS

Different cover crops benefit soil in different ways. Choose a crop that matches your needs or purchase a mix to maximize diversity. The crops listed in the chart below are some of the easiest to grow and control, and they span a wide range of applications. More options along with growing information can be found through seed retailers. Notice that I haven't recommended any grasses other than oats, which are easily winter-killed. Grasses have intricate root systems that can be difficult to control and remove without tilling.

COVER CROP	USE	COLD HARDY	PERENNIAL/ ANNUAL	WHEN TO PLANT	KILL METHOD
Buckwheat	Suppress weeds, loosen clay, increase phosphorus, generate biomass quickly, and attract pollinators	No	Annual	Anytime during summer and at least 8 weeks before a killing frost	Chop and drop or winter-kill
Oats	Build biomass; plant along with peas to provide them with support	Up to about 15°F (-9°C)	Annual	Late summer and early fall	Winter-kill
Field peas	Increase soil nitrogen and build biomass; most often planted with oats	Up to about 15°F (-9°C)	Annual	Late summer and early fall	Winter-kill or chop and drop
Tilling radishes	Break up clay, reduce soil compaction, deliver organic matter as deep as 30 inches below the surface, and draw nutrients close to the soil surface	Up to about 20°F (-7°C)	Annual	Late summer	Winter-kill or smother
Hairy vetch	Serve as fodder crop for livestock in winter, increase soil nitrogen	Winter hardy in many climates	Perennial	Late summer and early fall	Pull by hand
Clover	Serve as excellent living mulch (see page 131), add nitrogen, attract pollinators	Different varieties with varying hardiness up to 15°F (-9°C)	Annual and perennial varieties	Spring through early fall	Winter-kill or smother

protects the soil from harsh elements and heats the soil earlier in the season. Beneath the tarp, weed seeds are tricked into sprouting by the warmth, then they die from lack of light.

Perennial cover crops are less commonly used because they become a permanent garden fixture (or at the very least, one that is hard to remove). Plantings of vetch or perennial clover might be used for erosion control, as ground cover, as a living mulch, or in an area of a yard or garden with soil that needs intense, multiyear improvements. Some growers even plant perennial clover and other nitrogen-fixers in their walkways to improve localized nitrogen levels. Plant a perennial cover crop only when you don't plan to remove it after a single growing season.

DESIGN A CREATIVE EDIBLE LANDSCAPE

Sheet-mulched beds and hügelkultur beds can be constructed in irregular shapes filled with a mix of flowers, herbs, and edibles. The exact shape and size of the beds will depend on your available space. Curvy borders give the garden a soft, flowing appearance; straight borders feel more organized.

Create beautiful paths and borders. Rocks, logs, wood, or other aesthetically pleasing materials can be placed around the borders and used in walkways to enhance visual appeal. Combining a mix of materials adds texture and color to the garden.

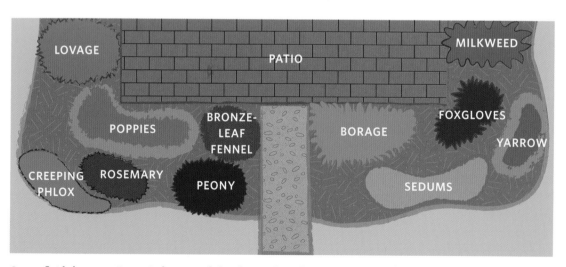

Create fluid shapes or "zones" of perennials by placing the tallest perennials in the back and the shortest in the front. Leave open spaces that can be filled with edible or flowering annuals to provide more texture and color in the garden.

Mixed beds filled with flowering perennials and annuals, peppers, cucumbers, bronze fennel, determinate tomatoes, and fruiting shrubs make excellent accents to a home or landscape.

Logs, rocks, and log rounds were placed as bed borders before the bed was filled with sheet-mulching materials. Six- to 8-inch-thick log rounds created a walkway through the center of the bed.

Edibles with colorful leaves such as beets and rainbow chard can be plugged into empty spaces in mixed flower beds to provide color in a unique (and delicious) way.

Larger beds are often better. Think big when it comes to designing mixed perennial beds. Large beds will fill in over time and boast a range of colors throughout the season. Smaller beds will be limited to just a few plants that won't necessarily provide continuous color.

Always follow recommended spacing. Perennials spread (some rather quickly), so it's important to allocate enough space for the plants.

Create "zones" or clumps of different plants. A well-landscaped bed contains few, if any, straight lines. Instead, plants are planted in fluid drifts throughout the beds.

Plant in odd numbers. Florists and professional landscapers frequently present a finished arrangement or bed using odd numbers to improve visual appeal. You'll need at least three of each type of plant to create a clump or zone. Five or seven are better when working with a large space, especially if the plant is small. You can get away with just one when that plant is particularly large, such as a Mexican bush sage, borage, false indigo, or rosebush. You may also want just one plant if you're filling a small bed.

Place the tallest plants in the back. Then set medium heights in the middle and the shortest specimens in the front. Creeping ground covers should be placed as far forward as possible. A taller plant can go in the center of a bed if it's to be viewed from all sides.

Leave gaps for annual flowers or vegetables. Gaps between perennials can be planted with either annual flowers that bloom all summer or colorful edibles like rainbow chard, red cabbage, speckled lettuces, purple kales, Romanesco, purple cauliflower, or beets. Tall plants like tomatoes, peppers, or red-leafed okra can add varying heights, colors, and textures, especially once fruits begin to ripen.

PICK OUT SOME PLANTS

The following lists are designed to help you get started with some of the most common perennials. These lists are by no means comprehensive, but all the plants should be readily available in most areas. Talk with local growers, shop at local nurseries and farmers' markets, and search online nurseries for some of the lesser-known edible and ornamental perennials. Most of these are planted like any other plant; those with specialized instructions are noted.

plants you can eat

- Asparagus (see page 128)
- Bloody dock
- Borage
- Bronze fennel
- Garlic (see page 122)
- Horseradish (see page 132)
- Lavender
- Lovage
- Mint
- Nodding onions
- Oregano
- Perennial onions (see page 126)
- Rhubarb
- Rosemary
- Shallots (see page 126)
- Sorrel
- Strawberries (see page 130)
- Sunchokes (see page 132)

plants that attract pollinators

- Asters
- Bee balm
- Black-eyed Susans and other rudbeckias
- Blanketflower
- Catnip
- Columbines
- Comfrey
- Coralbells
- Coreopsis
- Crocuses
- Daffodils
- Daisies
- Delphiniums
- Foxgloves
- Gladiolas
- Hyacinth
- Hyssops
- Joe-Pye weed
- Liatris
- Lilies and daylilies
- Lupines
- Milkweed
- Phlox
- Poppies
- Salvias
- Solidagos (goldenrod)
- Tulips
- Yarrow

KNOW AGGRESSIVE PERENNIALS

Perennials can be aggressive in one of two ways. Either they spread through underground root systems or they prolifically reseed. Some do both. A plant's ability to spread and take over is climate specific. Warmer climates often host a greater number of invasive species than cooler climates, so use caution when selecting plants for your garden. A quick internet search can yield a list of invasive plants in your area.

Creating a comprehensive list of aggressive perennials is nearly impossible because the success of a particular species is often region specific. It's also species specific. For example, Appalachian bellflower (*Campanula divaricata*) is native to North Carolina. Creeping bellflower (*C. rapunculoides*), originally from Europe, is listed as a noxious weed. I've provided a short list of potential problems to get you started, but always do your research before planting a new or unfamiliar perennial.

common aggressive perennials

- Butterfly bush
- *Campanula* (some varieties)
- Gooseneck loosestrife
- Horseradish
- Ivy
- Lamb's ears
- Lamium
- Lily of the valley
- Mints
- Nandina
- Obedient plant
- Oxeye daisy
- Raspberries
- Ribbon grass
- Snow-on-the-mountain (a.k.a. bishop's weed)
- Sunchokes
- Tansy
- Water hyacinth

Leaves of bloody dock are best eaten when young and tender in early spring. Its bright, lemony flavor makes an excellent addition to raw salads.

Don't mingle aggressive perennials with poor competitors. Aggressive perennials such as mint or sunchokes (see page 132) can completely take over and are therefore best planted alone in a bed with borders that are mowed. Asparagus is an example of a poor competitor that can quickly be swallowed up in a mixed perennial bed; it should be planted by itself.

Contain aggressive perennials in large pots or isolated beds. Ideally, these isolated beds will have a physical barrier (such as raised bed edging) and borders that are mowed to help prevent spread.

Dispose of aggressive perennials properly. Burn unwanted roots that are removed from beds or pots. If burning isn't an option, spread out the roots to dry in the sun for about a month. Never throw them in a compost pile or toss them in the woods where the roots can take hold.

NATIVE PLANTS BENEFIT US ALL

>>> Plants native to your region have a few advantages compared to common ornamental and landscaping plants. Natives . . .

- are some of the easiest plants to grow because they're well suited for local soil and climate conditions
- provide an important food source for local wildlife such as birds
- attract and support local pollinators that can improve fruit production in the garden
- rarely become invasive in their home region
- preserve local heritage and contribute to important genetic pools

Search the web for a list of native plants for your region. Many natives like this blanketflower serve as important food sources for local pollinators while beautifying the garden.

COST-SAVING STRATEGIES FOR SOURCING PERENNIALS

A trip to a garden store or farmers' market proves that the cost of perennials can quickly add up to hundreds of dollars. I use a handful of cost-saving strategies to build my garden diversity more affordably.

Ask your friends. A great way to acquire perennials is by gathering them from friends and neighbors. Most perennials readily spread and multiply, which means that most gardeners with well-established perennial beds always have more than enough plants to share. Perennials need to be divided every few years if they're to continue producing prolific blooms (see Dividing Perennials, page 134). Check with friends, neighbors, and local Facebook groups or garden clubs to learn when free perennials become available.

Buy bigger plants. A single large perennial can sometimes be divided into three or more plants. Select plants with the greatest number of stems growing out of the soil. Plants with just a few stems are too young for division, but multistemmed plants in large containers (7-inch pots and larger) with vibrant root systems can be divided. Avoid dividing a plant when it's in full bloom or covered in buds, or during midsummer, as disturbing the roots at these times can heavily

COLD STRATIFICATION CAN ENCOURAGE GERMINATION

>>> Some seeds require cold stratification—a period of prolonged cold—to germinate. This is particularly common for native species and plants whose seeds have a hard, protective covering. The covering delays germination until after a period of chilly but above-freezing temperatures. This prevents seeds from germinating during a temporary warm spell in late winter or early spring when the seedling could succumb to frost. A prolonged period of above-freezing chill lets the seed know that spring has arrived and it's safe to emerge.

Seed packets often indicate whether cold stratification is a requirement. Plants that require this include lupines, milkweed, St. John's wort, pincushion flowers, wild geraniums, some rudbeckias, larkspur, and lavender. Their seeds will sometimes germinate without cold stratification, but germination and early growth are improved with it. Direct sow these seeds outdoors in fall to naturally induce stratification. Alternatively, coax the seeds to germinate in spring by following a few simple steps.

1. Soak the seeds in water for a couple of hours.
2. Collect the seeds by straining them through a paper towel or coffee filter.
3. Transfer the seeds to a damp paper towel or coffee filter and spread them out in a single layer. The seeds can become moldy if the paper towel or filter is too wet.
4. Place the paper towel or filter into a ziplock bag, seal the bag, and store the seeds in the refrigerator for one month.
5. Plant the seeds immediately if they begin to sprout or when the one-month period is complete.

Grow them from seed. Some perennials are easily started from seed, while others prove more challenging. I've had much success with delphiniums, foxgloves, daisies, rudbeckias, yarrow, borage, sorel, feverfew, chamomile, gauras, bronze fennel, salvias, hyssops, and catnip. Lilies, irises, gladiolas, and other bulbed plants have proven slightly more difficult. Be sure to read the seed packet for details regarding germination requirements. Some perennial seeds can take months to germinate, while others require cold stratification (see Cold Stratification Can Encourage Germination, page 121).

Not all perennials bloom their first year while their roots are developing. Your investment in patience will pay off in the long run: You can have as many perennials as you want for the cost of a single packet of seeds and some potting soil. But don't be surprised if you must wait a bit for that perennial to be as prolific as your purchased plants.

Keep an eye out for large bags of spring and fall planted tulips, daffodils, and other bulbs as a cost-effective way to add early spring color.

stress the plant. If you acquire the plant at an inappropriate time for division, place it in the ground, allow it to establish itself, and instead divide the plant in fall or early spring.

GROW LOTS OF GARLIC

Garlic is one of my favorite plants to grow. The flavor of homegrown garlic is unmatched—and I'm not talking just about garlic cloves, which are the most commonly consumed part of the plant. Garlic leaves, flowers, and bulbs provide unique culinary contributions that can be enjoyed through spring and early summer.

Your climate will dictate which type of garlic you're able to grow. Hardneck varieties thrive in colder climates (permafrost is the only thing that can stop them) but fail to produce in warmer climates. The opposite is true for softnecks: They're better suited for warmer regions. Each variety has a specific USDA Hardiness Zone range, so you'll

Hardnecks (left) have an obvious stalk in the center of just one or two whorls of cloves. The stalk is the remnant of the scape. Softneck heads (right) are composed of a larger outer whorl of cloves with smaller cloves in the center and no scape.

need to ensure you've selected one that's suitable for your zone.

Hardneck garlic has a stem growing through the center of the bulb and a single whorl of cloves surrounding the stalk. Hardneck cloves have a stronger flavor, are easier to peel, and produce a scape—a flowering stalk—that can be harvested and eaten before the head is fully mature. Hardnecks can be stored for up to six months, variety dependent.

Softneck garlic lacks a center stem (thus, it doesn't produce a scape) and has numerous cloves per head, often with a milder flavor. The stems of this garlic can be braided together for storage because there's no firm center. Softnecks have longer storage life (up to 12 months, variety dependent) and can be consumed before maturation as green garlic.

HOW TO HARVEST

Use the scapes. In spring, hardneck garlic produces a false flowering head called a scape. If left to mature, the head will reveal a cluster of bulbils, the same reproductive structure as that of walking onions (see page 68). Garlic bulbils mature into large heads in two to three years, thus the reason why garlic is grown from cloves that mature in just a single growing season. Clip or break off the curly scapes when they're approximately 18 inches long to push more energy into the bulbs. Scapes can be eaten raw in salads; grilled, sautéed, or roasted whole; fermented; dehydrated; or added to any dish calling for garlic.

Harvest hardnecks and softnecks at different times. Harvest hardneck garlic when three to five bottom leaves turn brown and wilt and just four or five top leaves remain green. Harvest softnecks when the stems of three-quarters of the crop have developed a weak

Harvesting garlic scapes forces energy into the bulbs so they grow larger. Plus they're delicious!

spot directly above the bulb and the stems have fallen over. Harvest too early and bulbs won't have reached their mature size. Wait too long and bulbs will begin to deteriorate or start to sprout.

Remove garlic by grasping the stem a few inches above the soil line and firmly pulling it from the soil. You may need to carefully use a digging spade to loosen the soil if the stems break away from the bulbs as you pull them. Brush away any excess soil, but don't wash the heads.

HOW TO STORE

Garlic needs to cure (dry) for four to six weeks to improve flavor and storage life. Removing the stems or roots immediately after harvest creates a point of entry for bacteria and fungi. Garlic is therefore cured whole by drying it in a cool,

dark place such as a garage or basement. Hang bunches of 10 or spread out the bulbs (leaves and all) in a single layer on a drying rack or on top of cardboard to dry.

Softnecks can be braided together once the leaves have wilted and started to turn brown, about a week or two after being harvested (see Planting Perennial Onions, page 127). Hardnecks cannot be braided because of their hard stem.

The bulbs shrink just a bit and the leaves and roots become brown and brittle after four to six weeks. Once fully dry, brush off the dried soil along with any loose, papery outer covering. Clip off the leaves about an inch above the bulb and remove the roots at the base. Toss the waste into the compost. Bulbs should be stored in a cool, dark location with good airflow, such as a mesh bag in a cupboard or pantry or inside a cardboard box. Never store garlic in plastic bins or bags or in the refrigerator.

I harvest 300 to 350 heads of garlic every summer, which is enough to last an entire year and to seed the following year's crop.

GROW YOUR OWN SEED GARLIC

>>> Crops that are propagated vegetatively such as garlic and potatoes are more susceptible to disease than crops grown from seed. Disease is stored in the plant tissues and reemerges on the following year's crop. Crop rotation combined with proper curing and storage help reduce the spread of disease.

Grow about 20 percent more garlic than you'll eat each year if you plan to use a portion of the harvest to start the following year's crop. I plant between 300 and 350 cloves in approximately 200 square feet of gardening space, which provides plenty for eating and to seed the following year's crop.

When growing more than one variety, separate the varieties and label them well, as different varieties mature at slightly different times. At the very least, don't intermingle hardnecks with softnecks. Save your best and largest heads as seed.

Planting Garlic

Garlic bulbs are planted in fall and early winter when colder temperatures prompt root growth. Bulbs that are planted too early can succumb to fungus or bacteria while they sit underground waiting for a cold spell. Or they can sprout prematurely and experience frost damage in winter. Garlic that sprouts early will most likely survive, but it's best to time your planting so that the cloves don't sprout until spring.

If you miss the fall planting date, place the garlic in the ground as soon as the soil can be worked in spring. In this situation, hardnecks will likely be smaller and underdeveloped, but spring-planted softnecks may develop correctly. The reason? Hardnecks require a longer and more intense period of cold to stimulate growth and clove development.

1. Break seed garlic heads into individual cloves, leaving the paper on each clove intact. Only plant the largest cloves (eat the rest of them!).

2. Use a hand trowel to plant each clove 4 inches deep with the blunt side facing down, pointed side facing up. Space cloves 8 inches apart with 8 inches between rows. Aerate the soil with a digging spade (see page 62) prior to planting if the soil is heavy or compacted. Garlic prefers loose, fertile soil.

3. Cover the newly planted garlic with an inch of compost followed by an inch or 2 of aged wood chips or straw.

4. Keep the beds weeded. Garlic won't produce well if it has to compete for resources. Pull weeds when they're small; removing large weeds will disturb the garlic's roots.

GROW PERENNIAL ONIONS

Onions aren't often thought of as perennials, but there are so many perennial varieties—walking onions, shallots, and potato onions, to name a few—that I rarely plant the common annual versions found at grocery stores. Shallots and potato onions are smaller than a white or red onion, but they store well (up to 18 months for small potato onions) and don't require purchasing seeds or sets every year as long as you grow enough to seed the following year's crop. As with garlic, the best of the harvest is broken into individual bulbs and planted in late fall or early spring.

HOW TO HARVEST

Harvest the greens sparingly. Onion and shallot greens can be used as green onions but should be harvested selectively, as the leaves provide energy to the bulbs.

Remove the scapes. Not all perennial onions will produce scapes (flowering stems), but these should be removed to allocate more energy toward bulb formation.

Harvest when leaves have withered or fallen over. Potato onions are ready to harvest when the leaves develop a weak spot a few inches above the bulb and fall over. Shallots are ready when their leaves have fallen and started to dry out.

HOW TO STORE

Spread onions and shallots in a single layer in a warm, dry location. Potato onions require about three weeks until the leaves feel dry and the outer skin becomes papery and tight around the bulb. Shallots require one to two weeks. After curing, braid the stems together or remove the dried stems and roots and store the harvest in a breathable bag or basket in a cool, dark location.

Braiding shallots and softneck garlic is a handy way to store them.

Perennial potato onions are planted and grow much like shallots but the resulting bulbs are often larger. Bulbs are 1 to 3 inches across and can be harvested as immature green onions or when the leaves fall over directly above the bulbs.

Planting Perennial Onions

In many parts of the country, shallots and potato onions can be planted twice per year—once in late winter and again in early spring as soon as the soil can be worked. The two plantings will mature at nearly the same time, but the yield is said to be two to three times higher with fall plantings. (Growers below Zone 4 must wait until spring to start their onions, because fall-planted bulbs are unable to survive harsh winters.)

The smallest bulbs are often planted in early spring, and the largest planted in fall.

1 Top-dress the bed with compost and use a digging spade to loosen compacted soil. Like garlic, onions are heavy feeders, and bulbs won't properly expand if planted in an area with compacted soil.

2 Break any clumps of onions or shallots into individual bulbs.

3 Use your hand or a trowel to place the shallots and potato onions about 1 inch deep with the blunt side down and pointed side up. Space shallots and onions 9 inches apart with 1 foot between rows.

4 Cover the bulbs with 1 to 2 inches of aged wood chips or straw to retain moisture and prevent weeds. In colder climates, up to 8 inches of mulch (2 inches at the very least) can be applied as cold protection for fall-planted bulbs. Gently remove the mulch in early spring before the bulbs begin to sprout. Unlike garlic, onions don't like to be buried deep.

Shallots create a perfect tuft of onion greens as they emerge in early spring.

ADD ASPARAGUS

You have to hurry up and wait for asparagus, but it will produce abundantly for up to 30 years if placed in a desirable location. The plants are heavy feeders and poor competitors, which means they need nutritious soil, lots of sun, and a solo spot in the garden where they can thrive undisturbed. Don't let asparagus get overrun by weeds, or you'll see an immediate decline in production.

Plant six to eight 1-year-old crowns. While it's possible to start asparagus from seed, that results in another year of waiting time. For that reason, most people plant crowns, which are clusters of year-old asparagus roots. A family of four can easily consume the number of spears produced by six to eight crowns. Plant more if you're hoping to preserve or share with friends.

Harvest three-year-old crowns. Allow one- and two-year-old spears to build energy for the plant's roots by maturing into tall, wispy, fernlike stems. The female plants bear red berries in summer.

Stop harvesting after six to eight weeks. In year three, harvest asparagus in spring by cutting off spears at the soil line when they're 10 to 12 inches tall. After six to eight weeks (I generally stop picking at the beginning of July), allow the stems to mature and build energy for the following year.

Remove the dead stems in winter. The tall, fernlike stems change from green to brown when temperatures consistently drop at the end of the growing season. Cut off the fronds at the soil line and add the dead material to the compost pile. Be sure to complete this step by the time your new asparagus spears sprout in spring.

Harvest asparagus spears as they emerge from the ground. When left to mature, they flourish into tall, wispy, fernlike stems.

STEP-BY-STEP

Planting Asparagus Crowns

Asparagus are dioecious, meaning there are male and female plants. Females produce beautiful (inedible) red berries in fall and generally grow larger spears. Males produce smaller, more abundant spears and are thought to live longer. However, there's no need to plant both; you can select any combination of male and/or female. In fact, many hybrid varieties, including 'Jersey Giant', are all male, so selecting females isn't even an option.

Asparagus crowns are best planted in spring. They'll be easier to plant in an established bed or a new sheet-mulched bed that is covered in ample compost because the crowns need to be buried deeply. This would be challenging, if not impossible, in a new hügelkultur bed with large pieces of organic matter directly below the surface.

1 Dig an 8- to 10-inch-deep trench, allowing for 1 foot of length per crown (that is, four crowns will need 4 feet). Space rows 2 to 3 feet apart to account for spread.

2 Line the crowns down the length of the trench, bud to tip. Each bud should be 1 foot from the next bud.

3 Apply 2 inches of compost directly on top of the crowns, followed by another 2 inches of soil removed from the furrow.

4 Water immediately and keep the crowns moist over the next few weeks while you wait for them to emerge.

5 Add more soil a few weeks later once the ferns have emerged and are a few inches tall. Continue to backfill until the soil from the furrow has been replaced.

THREE MORE CROPS TO CONSIDER

There are so many interesting crops to focus on when you're at the point of expanding your food garden. Garlic, onions, and asparagus have broad appeal and are fairly easy to grow, so I introduced them with detailed instructions. Here I briefly discuss a few others that are worth experimenting with.

STRAWBERRIES

Strawberries are among the first fruits to ripen in summer, and they produce reliably if they receive some annual attention. Strawberries are relatively poor competitors and perform best when planted alone. The plants do better when kept dry and therefore benefit from being planted in a raised bed with efficient drainage.

Some gardeners allow strawberries to creep in below other plants as a ground cover, but not everyone will have success with this method. Low-lying ground covers are prone to moisture issues, and strawberries often succumb to fungus as they ripen. Growers in hot, dry regions may have better success using strawberries as a ground cover than growers in a wetter climate. Mulching strawberries isn't always beneficial, either, as the mulch can retain moisture and attract slugs and other insects that will eat your fruits.

Strawberry plants reproduce vegetatively by shooting out new plants via runners every year. Ideally, gardeners thin strawberries every fall or winter by removing old plants and allowing the new ones to produce the following year. Production drastically decreases as the plants age, which means you'll have to continually remove older plants to allow space for the younger plants to thrive.

Older strawberry plants are often larger and have a somewhat woody stem. Remove the parent plant and replant the younger shoots in the late fall.

Sweet alyssum makes an excellent living mulch planted at the base of peas, tomatoes, cucumbers, and other vertically grown plants. Its compact growth habit protects the soil while producing prolific purple or white flowers that attract pollinators.

LIVING MULCHES

>>> Living mulches are low-growing plants that are placed below taller plants or along borders to provide diverse benefits. Ground covers like hens and chicks sedum protect the soil from the sun and retain moisture. Matting thymes prevent erosion and can even handle some light foot traffic when planted in walkways. Sweet alyssum planted along the borders of trellised cucumbers attracts pollinators. Clover planted beneath tomatoes adds nitrogen to the soil while preventing water from splashing onto the tomato leaves and spreading blight.

Growers in arid climates can try using strawberries as a living mulch beneath blueberry bushes and other tall plants, but this technique won't work in climates where the plants remain too moist from being shaded.

Plant annual cover crop seeds or place annual plants once the main crop is at least 10 inches tall (2 feet for slow-growing plants like peppers). Keep the seeds well watered until the seedlings are established. Give the living mulch a trim or remove it entirely if it ever looks like it will outcompete the main crop. Leave perennial living mulches in place year after year. Clear out openings in the ground cover for new plantings.

Sunchokes can be roasted, sautéed, or sliced thinly and eaten raw in salads.

Horseradish is a lush, full plant with two different leaf shapes—one lobed and one not. Thick, large roots develop as the plant ages, but numerous gnarly roots from younger offshoots such as the one shown here are plentiful.

SUNCHOKES

Sunchokes, or Jerusalem artichokes, produce a sunflower-like bloom and an edible tuber. Isolate sunchokes in an area with a mowed perimeter because they spread so aggressively. Plant sunchoke tubers about 6 inches deep in early spring and wait 110 to 150 days to harvest. Remove some of the flowering stalks as they emerge to help push energy into tuber production. Harvesting tubers after a frost can improve their flavor.

Unlike potatoes, sunchokes can't be cured for long-term storage. Harvest tubers starting in late fall and continuing through early spring if the ground doesn't freeze. Consume freshly dug tubers within two to three weeks. Allow 20 percent of the tubers to remain underground to produce more sunchokes the following season.

HORSERADISH

Horseradish is often grown in large pots to avoid unwanted spread. Plant horseradish roots in early spring and delay harvesting until its second year of growth. Harvest in fall by digging up a 12- to 18-inch-deep trench along one side of the plant. Remove side roots that are visible from the trench, trying to leave the main root intact. However, horseradish is far from sensitive. Digging up an entire plant and replanting just a few side shoots works perfectly fine, too.

MAINTAINING HEALTHY PERENNIALS

Perennials require very little attention after establishment. Keep perennial beds mulched to build soil, retain moisture, and suppress weeds. Deadheading perennial flowers can sometimes help lengthen bloom time but not always as effectively as it does with annuals. Many perennials will likely only bloom for up to eight weeks, regardless.

To control their spread and to encourage prolific blooms, divide perennials every three to four years. Overcrowded plants won't produce nearly as many flowers as those that receive adequate space.

Remove dead plant material in fall to improve garden aesthetics, prevent the spread of fungus, and eliminate habitat for undesirable insects and slugs. However, allowing spent foliage to remain and overwinter in place helps protect the roots and soil from temperature extremes and can provide winter protection for beneficial insects. For these reasons, there's no right or wrong time to clean up the garden as long as spent foliage is removed before next year's foliage sprouts. A few plants such as poppies, irises, daylilies, and daisies begin coming up in late winter in some climates. In those cases, be sure to remove the previous year's foliage before new growth begins.

TIPS FOR HARVESTING EDIBLE PERENNIALS

 Perennials should be selectively harvested to maintain energy for years to come. Overharvesting can impede a plant's growth or reduce the stock so much that yield is inadequate.

Tips for Harvesting

- Perennials that sprout from the ground can be harvested entirely. Oregano, mint, chives, walking onions, bloody dock, sorrel, and other perennials that sprout from the ground every spring can be chopped off at ground level midseason and allowed to resprout.
- Woody perennials need selective harvesting. Thyme, rosemary, lavender, and sage form woody branches that produce new growth every spring. Harvest sprigs as needed, preferably above a bud so that the plant can bush out and grow fuller. Never harvest more than half of a woody perennial at a time.
- Harvest most roots in late fall through winter. Horseradish and sunchokes mature over the summer for a late-fall harvest. Roots can be left in the soil and harvested throughout winter on an as-needed basis in climates where soils don't freeze.
- Young leaves are most tender. Harvest leaves at any point in the season, but it's best to select young, tender leaves. This is especially true for lovage, sorrel, bloody dock, and other perennials whose leaves become fibrous and sometimes bitter as they mature.
- Harvest rhubarb in spring. Pick rhubarb stems as soon as the leaves and stems are fully mature (toss the poisonous leaves!). Stop harvesting once the plant begins to flower.

Dividing Perennials

Divide a perennial in late fall to allow time for the roots to establish themselves before spring. If you miss the fall window, divide plants during winter or in early spring immediately after they start to emerge from the ground. Never divide a perennial when it's covered in buds or is flowering, or in the heat of summer.

1   Start by cutting the foliage down to 2 to 6 inches. You'll be able to more easily see how to divide the rootball. Even better, dig the plant in the winter when it has no foliage at all.

2   Dig a circle around the plant, about 4 to 6 inches away from its base, using a shovel or digging spade. Pry the rootball free from the soil using the shovel or spade. Try to keep the soil intact to avoid destroying small feeder roots. Remove the rootball from the ground.

3   Inspect the roots to determine the best way to divide the perennial. Iris rhizomes or daffodil bulbs, for example, are best divided by breaking or cutting the root systems apart using your hands and a sharp knife or pruning saw. Tangled clumps of roots such as those found on daylilies, daisies, and most other perennials are divided using a shovel or a saw.

4   For especially dense rootballs, place the rootball on the ground so that it won't roll, then swiftly push a shovel through the rootball and break the clump into three or more chunks to be replanted.

5   Replant the desired number of clumps, being sure to allow plenty of space between clumps. Mulch around the newly planted perennials and water well for the next few weeks while the plants reestablish their roots.

Divide daylilies and other herbaceous flowering perennials every three to five years to encourage prolific blooms.

5

four-season
GROWING

In most climates, you can help cold-hardy vegetables thrive during winter by using a few simple tricks to extend the harvest season. You won't need to preserve cold-hardy vegetables like kale, collards, leeks, beets, and lesser-known greens like chicory and cress when you can grow them year-round. Instead, you can focus your preservation efforts on fleeting summer foods like berries, fruits, tomatoes, tomatillos, squash, cucumbers, beans, and okra.

Start your winter growing experiment with a low-cost option such as a cold frame, which can extend your salad season by several months.

the plan

- **EXPAND THE GARDEN BY ANOTHER 400 SQUARE FEET** to accommodate extra crops for year-round growing. Plants to be harvested during winter and into early spring must be planted in late summer and fall when many summer plants are still thriving and producing. You need more space.

- **COLLECT MATERIALS FOR NEW INFRASTRUCTURE.** You'll need row cover (an insulating fabric that helps keep plants warm and protected) or rolls of clear plastic, as well as supporting wires or plastic hoops. If you decide to build cold frames or a greenhouse, you must gather those materials, too.

- **CREATE STRAIGHT, NARROW BEDS.** You'll install low tunnels in the new beds to protect the vegetables from winter temperatures (see page 143). These are most easily installed atop long, rectangular beds that are 2 to 3 feet wide.

- **CHOOSE VARIETIES FOR WINTER GROWING.** Some varieties of spinach, kale, collards, cabbage, and turnips, for example, won't even need insulation in milder climates. I'll also introduce a few new greens—radicchio, chicory, cress, endive, chickweed, and frisée are some of the hardiest options available.

Greenhouse design is infinitely variable. Mine is 315 square feet but a smaller one may serve your needs just as well.

BEFORE YOU DIG IN

As with all new gardening endeavors, learning to properly manage a winter garden takes practice. I'll offer plenty of guidance but allow yourself time to get to know your individual challenges, familiarize yourself with varieties that work for your climate, and learn the techniques best suited to your space, energy levels, and desired outcomes. Be patient as you learn to successfully cultivate a winter garden.

Winter growing takes a bit more effort than summer growing because temperature fluctuations need to be managed by opening and closing enclosures designed to trap daytime heat from the sun. Some insulating enclosures become hot in full sun even when outside temperatures are below freezing. Not all the options allow rainwater to enter the soil, which creates watering challenges. Different options demand different levels of attention, meaning you need to decide how much money, time, and effort you're willing to invest in winter growing when choosing a setup.

Purchase and assemble materials by midsummer so that the actual growing can start in fall. I prefer to build infrastructure in early spring or late fall, when temperatures are pleasant for outdoor work and I'm not inundated with garden chores.

HARNESS THE GREENHOUSE EFFECT

Insulating plants with garden fabric, plastic sheeting, cold frames, or greenhouses traps heat from the sun during the day, heating the soil beneath these structures. The heat from the soil radiates back into the insulating structure at night to maintain temperatures slightly above ambient and to protect plants from heavy frosts. The challenge is that these structures are efficient at trapping heat when the sun is shining, but the heat is lost once the sun goes down, meaning plants experience major temperature swings between day and night.

My greenhouse can easily reach 75°F (24°C) on a sunny, 30°F (−1°C) day even with all the roof vents open. The vents close when the sun starts to go down, and the temperature inside the greenhouse quickly drops to 5° to 10°F (3° to 6°C) above ambient. That can be a nearly 70°F (39°C) temperature swing within a 24-hour period, and the effect can be even more pronounced for cold frames and unvented low tunnels covered with plastic sheeting.

The insulating options discussed here—cold frames, low tunnels, and greenhouses—trap and release heat with varying efficiency, so they must be managed with differing levels of attention. Be realistic about your willingness to oversee winter crops and choose a setup that coincides with your desired level of effort. Installing low-tech automated systems for use with cold frames and greenhouses can make management far easier, but the initial cost investment will be higher. Learn about all the options and choose a method that feels right for you.

CROP ROTATION BASICS

>>> Perfecting crop rotation in the home garden can be challenging, as many of the crops grown in a diverse home garden represent a few common plant families. You can approach crop rotation in two different ways. The more in-depth method requires that you try not to plant anything from the same family in the same spot for at least three, ideally four, years. The easier method (which is how I generally approach crop rotation) is to focus on rotating a few specific crops that include heavy feeders, plants prone to disease, and nitrogen fixers and let the rest work itself out.

ROTATE PLANT FAMILIES

- **Alliums**: garlic, onions, shallots, and leeks
- **Betas**: beets and chard
- **Brassicas**: cabbage, collards, kale, radishes, mustards, kohlrabi, broccoli, turnips, Brussels sprouts, and cauliflower
- **Cucurbits**: cucumbers, melons, pumpkins, gourds, and squash
- **Legumes**: green beans, dried beans, clover, and peas
- **Nightshades**: tomatoes, potatoes, tomatillos, eggplants, peppers, and ground-cherries
- **Umbellifers**: parsley, carrots, parsnips, dill, celery, and fennel

ROTATE SPECIFIC CROPS

- Beans
- Corn
- Cover crops
- Garlic
- Potatoes
- Shallots
- Tomatoes

ROTATE CROPS FOR SOIL HEALTH

Intensive crop rotation isn't often practical in small gardens because many of the most common crops represent just a few plant families (for instance, collards, kale, and radishes are all members of the genus *Brassica*). Replacing one crop family with another crop family doesn't always work out. As plant diversity, production, and available garden space increase, however, crop rotation becomes possible and extremely beneficial. Crop rotation is a great tool to add to your arsenal of soil management strategies. It prevents nutrient depletion and the spread of soilborne diseases and pests.

Furthermore, year-round production creates opportunities for more intensive growing. When you start experimenting with winter gardening, you'll likely find that you have vacant beds periodically. But as you master succession planting and growing seedlings in flats, you'll find that most of your beds have the potential to remain filled year-round.

This type of intensive growing benefits living soil in many ways. Roots stimulate soil processes throughout winter by feeding belowground decomposers and nutrient cyclers, helping to keep that essential community alive and well. Aboveground biomass protects the soil from the sun, prevents evaporation and excessive dryness, and insulates the soil from harsh winter temperatures. However, intensive growing, if not properly managed, can also deplete nutrient reserves and encourage the spread of disease.

When growing year-round, you may need to bump up the frequency of annual compost additions to twice per year for younger beds or those lacking adequate organic matter. Heavy feeders such as tomatoes, melons, potatoes, garlic, and squash deplete nutrients. Rotating them with nitrogen-fixing legumes such as peas and beans can help replenish nitrogen. Brassicas are useful to rotate because they're thought to help cleanse the soil. Tomatoes and potatoes spread soilborne disease. Marigolds help control nematodes, a belowground pest.

A crop rotation plan for a fixed trellis could be planting beans in year one, tomatoes in year two, and cucumbers or melons in year three.

A cold frame is easy to build from upcycled materials and can be any size that works for your space.

COLD FRAMES ARE SMALL AND SIMPLE

These low-cost mini greenhouses resemble raised beds covered with plastic or glass to trap heat from the sun. Cold frames are incredibly efficient and can be constructed from a variety of materials, including scrap wood, logs, straw bales, old windows, and/or clear plastic (be careful with glass, as a broken window can create a mess that's not easy to remedy).

The lid of a cold frame needs to be sturdy if it's to bear the weight of ice and snow. The lid can be positioned flat if you're using a strong, solid material such as tempered glass or clear fiberglass, but even thick plastic sheeting needs to be positioned at an angle so that rainwater can't collect on its surface and cause the center to sink and eventually collapse. Regardless of the material being used, orient the clear cover to face south to collect the greatest amount of sunlight, especially in winter when the sun is lower in the sky.

Cold frames require ventilation on sunny days to prevent plants from drying out or overheating. You can manually open and close cold frames based on weather predictions, but I highly recommend installing an automated hinge that's set to open when the cold frame reaches a certain temperature. I can't tell you the number of times I've forgotten to vent a cold frame before leaving for the day and have returned to a casket of dead plants. There are both electric and non-electric hinge options.

Nonelectric automated hinges designed specifically for greenhouses and cold frames have cylinders filled with wax or oil that expands as the temperature inside the greenhouse or cold frames rises. As the medium expands, it presses on the hinge and opens the window or louver. As the interior temperature decreases, the contents of the cylinder cool and contract and the window slowly closes. These hinges are simple to install, and the peace of mind they provide is well worth the minimal investment of $25 to $50.

LOW TUNNELS ARE VERSATILE AND MOBILE

Low tunnels protect crops from damaging winds and frost settling on leaves and provide a small amount of additional warmth. Build them no more than 3 feet wide and 18 to 24 inches tall to best trap heat. Besides, few winter crops grow taller than 18 inches.

CHOOSE YOUR SUPPORTS

Wire hoops or plastic pipes (PEX tubing, PVC, or even cut-up Hula-Hoops) can be bent into half circles across the tops of beds and covered with polyethylene (clear plastic) or the breathable garden fabric known as row cover. Plastic hoops can be permanently fastened over raised beds using metal brackets, or the ends can be slid over pieces of rebar that have been driven into the ground.

Low tunnels are easily converted from providing insulation with row cover to protecting against pests with insect netting.

Plastic hoops can be permanently fixed over the top of raised beds using brackets.

Insert 6 to 8 inches of the wire into the ground on one side of the bed, bend it over the bed, and insert the other end in the ground.

I prefer wire hoops because they're more affordable, last for decades, and are easy to move. Simply insert 6 to 8 inches of the wire into the ground, bend it over the bed, and insert another 6 to 8 inches on the other side. Wire hoops are available in several standard lengths, but 76-inch wires are most common. Plastic hoops made with PVC or PEX are more customizable as they can be cut to specific lengths. For this reason, plastic hoops would be desirable when trying to insulate a 4-foot-wide raised bed where available wire lengths would be insufficient.

Hoops are often spaced 6 feet apart. However, spacing should be based on local weather patterns. Heavy snows and winds can cause low tunnels to collapse. Closer spacing such as 2 to 3 feet between hoops offers better support.

CHOOSE YOUR COVERING

I'm discussing two choices when it comes to insulating low tunnels—polyethylene film (clear plastic) or row cover. Other covers exist, but these two are the most common and can be reused for four to five years if properly cared for. Both are available from greenhouse supply stores in large rolls of varying widths and lengths. Your chosen width will ultimately depend on the width of the beds and the length of the hoops, but 10-foot-wide covers work well for most home garden applications. You need enough to cover the hoops plus 12 to 18 inches of excess on both sides for pinning down the cover.

Clear polyethylene (PE) film is incredibly good at trapping the sun's heat. It's so effective that it'll require venting on sunny or warm days to prevent overheating. (Read: You'll be opening and closing the tunnels based on the weather.) The heat is quickly lost as the sun goes down, although it keeps plants about 5°F (3°C) warmer than ambient temps while offering protection from heavy frosts. Use 4- or 6-mil PE covers. Thinner films are available, but they tear easily and don't offer the same level of frost protection. PE film is waterproof, so you need a plan for watering, whether it's removing the film when it rains, setting up an irrigation system, or watering by hand as needed.

Row cover fabric doesn't get as hot. Row cover is available in varying thicknesses. Thicker fabrics insulate better but block more light. You're looking to strike a balance between light transmission and insulation. Start with a fabric that allows 70 percent light transmission. These fabrics will protect plants from frost down to 26°F (−3°C). Thinner fabrics offer less frost protection and may therefore be more desirable in warmer regions.

USE HOOPS WITH INSECT NETTING, TOO

>>> In spring and summer, flying insects such as moths lay eggs on crops that hatch into damaging larva (caterpillars). Caterpillars can be controlled using a low-toxicity organic pesticide, *Bacillus thuringiensis* (Bt), derived from soil-dwelling microbes. But you'll have to spray repeatedly on a schedule (like you'd do with any pesticide—every seven days in this case), as the bacteria kill only hatched caterpillars. Bt does nothing to protect from future infestations or eggs that have not yet hatched.

The better approach, therefore, is to keep susceptible crops covered with insect netting, a fine, breathable fabric that prevents moths from laying their eggs on crops. When it comes to organic growing, brassicas are the crops most susceptible to larval damage from cabbageworms. And fortunately, these low-growing crops can easily be protected with hoops and netting. Grow brassicas under an insulating cover in winter, and immediately replace the insulating layer with insect netting in spring.

Secure insect netting over hoops to prevent cabbage moths from laying eggs on brassicas. Place the netting in early spring before the moths emerge.

Drape the covering (row cover is shown here) along the length of the wires, pull it taut, and secure it in place using smooth rocks or sandbags that won't snag or tear the fabric.

Because the fabrics are breathable, venting is necessary only on warm days. And even better, rainwater can easily get through. For these reasons, row cover can be a set-it-and-forget-it situation for most of winter. Venting will be necessary in spring and fall when temperatures fluctuate between warm and cold.

Secure the covers along the base with sandbags, clothespins, or large rocks. Specialized clips for different systems are available if you want to spend the extra money. Avoid using objects that could snag the covers during installation or removal. Never use fabric pins that look like large staples. They're convenient to install but extremely difficult to remove, especially when frozen, and they poke holes through the fabric.

At the end of the season, allow the covering to completely dry. Neatly fold the PE film or row cover and store it out of the weather to increase longevity.

MORE USES FOR ROW COVER

>>> Row cover can be draped directly on the ground to help retain moisture when planting seeds such as carrots that have extremely finicky germination requirements or when planting seeds in the heat of summer when soils can quickly dry out. Remove the row cover once the seeds germinate.

It can also be used in winter to hold dried leaves and other insulating materials over root crops. Insulated turnip, radish, rutabaga, carrot, beet, and parsnip roots often survive winter even if the foliage suffers frost damage. Pile a foot or two of dried leaves on top, then tack down the leaves with the row cover.

A GREENHOUSE IS WORTH THE EXPENSE

A permanent, insulated structure offers more space and provides better year-round climate control, especially in the cooler months. Greenhouse frames can be made of wood, aluminum, PVC, plastic, or some combination of these materials. Glazings (the clear walls) are made from glass, solid polycarbonate panels, or polyethylene (PE) film or sheeting, all with varying longevity and ability to insulate.

You can certainly build your own greenhouse from new or repurposed materials and get as complicated as you'd like with the design. If you're like me, however, and lack the skill set, time, and energy to design your own, you'll need to purchase a kit. The following suggestions are, in most instances, applicable for both DIYers and those who are assembling kits.

Know your budget. Cost is usually the most prohibitive factor when selecting a greenhouse. You can expect to pay an additional 10 to 20 percent for related expenses such as installation labor, shelving, seed-starting mats, raised bed materials, pots, soil, compost, and so on. Be sure to select a greenhouse priced below your total budget in order to account for unexpected expenses.

Know your intended use. Are you interested in extending the season by a few weeks or months, or do you wish to keep tropical fruits alive during the winter? Heating (and cooling) a greenhouse can be especially pricey. Your intended use will therefore inform the size of the greenhouse and type of frame and glazing. Select a small, well-insulated structure for heating, but go as big as you'd like if you don't have plans to heat it.

A greenhouse is an incredible resource for starting seeds, hardening off started plants, and extending your growing season year-round.

Select a model with adequate ventilation. Ventilation is vital to regulate the temperature inside the greenhouse, improve air-flow, and prevent disease and pest infestations. Automated vents and windows make greenhouse management far easier, especially when major temperature swings occur during the winter. Solar-powered fans can improve air circulation on sunny days without needing electricity, but electric fans are better at moving large volumes of air.

If designing your own greenhouse, be sure to install plenty of windows, vents, or fans. Greenhouses made from bent aluminum pipes covered in PE film should be equipped with rollup sides. This additional cost and installation effort will be well worth the investment.

Choose the glazing. Greenhouses are most frequently insulated with glass, polycarbonate panels, or PE films and sheeting, each with its own benefits and disadvantages.

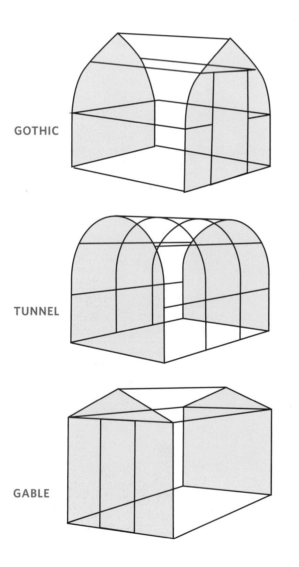

GOTHIC

TUNNEL

GABLE

GREENHOUSE SHAPES

》》》 Gothic, tunnel, and gable are the most common shapes for personal greenhouses, but there are many designs out there. A lean-to or dome might work better than one of these for your situation.

A gothic shape is best at shedding snow and has relatively high sidewalls for ease of access, making this one of the more popular designs. However, bending a flat polycarbonate panel into place during con-struction is difficult.

Tunnels are the most common shape for structures covered with PE film and sheet-ing. Their sidewalls are extremely short where the walls meet the ground, so you'll have to be mindful of bed and plant place-ment; you won't be able to walk through this area.

Gables have good sidewall height, which improves layout versatility, but they don't shed snow as readily as some of the other shapes.

Glass is the most expensive option if using new materials but is the longest lasting (if the panels don't break). Single-paned glass is an extremely poor insulator, so it's best to use double-paned. Use glass only in a location that is likely to be free of falling branches and other debris.

Polycarbonate panels cost less than glass, are up to 100 times stronger, last a minimum of 15 years, and come in varying thicknesses. Consider paying more for 16 mm triple-wall polycarbonate panels, which offer better insulation than 8 mm double-wall panels. A UV-protective coating can prolong the life of the panels.

PE films and sheeting are the least expensive options but need to be repaired occasionally and should be completely replaced every four or five years. PE film offers the best insulation for the price, but PE sheeting is thicker and has additional features such as an option for creating an inner chamber that improves insulation.

USING A KIT

When choosing a kit, read the reviews and look to see who has published the review—is it the manufacturer or a third party? Kits often receive a wide range of both good and bad feedback. Pay attention if you see the same complaint repeatedly.

Download the assembly instructions for the kit prior to purchasing and check for proper grammar and understandability. Many of the kits are manufactured overseas; some come with poorly translated instructions. Also search the internet for instructional videos to accompany the kit and to verify that the assembly instructions are complete, organized, and easy to understand. Even the best written instructions can benefit from video supplementation.

A FEW MORE TIPS

Orient the greenhouse east to west. You want to capture as much light as possible, especially in winter. The long side of a greenhouse should face the path of the sun and not be shaded by existing buildings.

Greenhouses benefit from summer shade. Shade from a building is too intrusive, but partial shade from deciduous trees can keep a greenhouse slightly cooler in summer months and prevent the need for shade cloth. Because deciduous trees lose their leaves in fall, more light and heat enter the greenhouse in winter. If natural shade isn't an option, consider using shade cloth in summer.

Build a proper foundation. You're fitting perfectly rectangular panels together or covering

Eight-inch concrete forming tubes were fitted tightly into 24-inch-deep footers and used to level the aluminum base with a sight level.

This large hügelkultur bed incrementally raises the temperature in my greenhouse as it decomposes.

better insulate the building). Replacing a rotten foundation can be extremely difficult. Leveling an aluminum base is more challenging but will be worth the effort in the long run.

Anchor the structure to the ground. Greenhouses are extremely light for their size and can be blown about like a giant kite if not fitted with a proper foundation. In most cases, you'll have to install concrete footers or drive metal posts into the ground below the freeze line. The larger the structure, the heftier the anchoring system. Follow the kit's instructions on how to best anchor the greenhouse.

GENERATING EXTRA WARMTH

Use thermal mass to create more warmth in winter. Water has a high heat-holding capacity and can therefore be used to trap heat from the sun during the day and release that warmth at night. One technique is to paint large barrels black, place them along the sunniest side of the greenhouse, and fill them with water to collect heat. Be aware that this technique takes up a great deal of space (and a lot of barrels are needed for large spaces) if it's to radiate significant amounts of heat. Lining walkways with bricks, concrete, and other heat-absorbing materials is another way to aid in daytime heat collection.

Compost naturally generates heat as it decomposes, and some growers place compost heaps inside their greenhouses to capture a bit more heat.

To maximize heat absorption in my greenhouse without losing space to barrels or compost heaps, I built large raised beds (the largest is positioned along the sunny side of the greenhouse to collect heat) and filled them with loads of organic matter including wood, sticks, leaves, chicken manure, cardboard, and other materials

the structure in perfectly rectangular PE film. These tasks become nearly impossible if the foundation was improperly installed. The base needs to be level, square, and plumb. Don't scrimp on this vital step. Plan for the additional cost if you don't have the experience and equipment to do the job yourself.

Think twice about a wooden base. Polycarbonate kits usually come with an included or optional aluminum base. Squaring and leveling a wooden base is far simpler, hence why most people choose to first build a wooden base and then secure the aluminum to the wood. The problem is that even pressure-treated wood will rot over time when in contact with the ground (greenhouses require ground contact to seal and

you'd use when building a hügelkultur bed. Additionally, I covered the entire floor with 12 to 18 inches of wood chips. The decomposition of the contents of the beds and the wood chips generates small amounts of heat, while the mass of these additions captures even more heat. The impacts are small but effective.

PREPARE FOR HEAVY SNOW

Winter harvests are possible in most cold and snowy regions with a bit of extra protection. Cold frames, low tunnels, and greenhouses need to be designed to hold an appropriate snow load.

Promptly remove heavy snows to prevent structure collapse and to let in light. Smaller snow loads can be left in place during cold snaps as a protective, insulating layer.

Space low tunnel hoops closer together (2 to 3 feet) to prevent collapse under heavy snows. Doubling up on winter growing techniques such as using two layers of row cover, installing low tunnels inside a greenhouse, or building double-walled cold frames provides even more protection to keep winter crops alive and well. Plant selection will be especially important, as you'll need to grow the hardiest varieties you can find. Don't be afraid to experiment until you find what works!

Clearing the snow from a greenhouse in the morning allows more light to enter and quickly heat the structure for the day.

Matthew and Heather Kasvinsky built this greenhouse on their homestead in South Duxbury, Vermont, using four 4 × 16-foot panels for a final footprint of approximately 112 square feet. Photo courtesy of Heather Kasvinsky.

LOW-COST, DIY CATTLE PANEL GREENHOUSE

>>> Large cattle panels (the same fencing recommended for use as a trellis in Step 1) can be bent into a horseshoe shape, attached to a solid wooden base, and covered with PE film to create an affordable greenhouse. Connect as many as four 4 × 16-foot panels to create a greenhouse that suits your needs. (Note: the panels come in widths ranging from 36 to 52 inches. For simplicity, I'm referencing 48-inch panels, but be sure to measure your panels before beginning construction.)

The end walls of cattle panel greenhouses are easily vented with doors and windows, but venting the side walls is harder. Building a greenhouse with more than four panels (longer than 16 feet) will require ventilation along the length of the structure or installation of a powerful fan, because end wall vents alone won't adequately handle a structure longer than 16 feet. Search the internet for plans and instructional videos on how to construct a cattle panel greenhouse ranging from simple to complex.

This 'January King' Savoy-style cabbage survived without protection for an early-winter harvest. Savoy (crinkly) leaves are often an indicator of superior cold tolerance when compared to smooth leaves. This is true not only for cabbages but for kales and other brassicas.

Many winter crops, including these 'Uzbek Golden' carrots, become sweeter after a few frosts. Note that although many carrot varieties never grow to the size of carrots found in the grocery store, the homegrown crops taste far superior.

CHOOSE WINTER VARIETIES CAREFULLY

Growing just any cabbage in winter isn't possible in all climates. In Zone 7a, I have to find the most cold-tolerant variety I can and keep it insulated under row cover or in the greenhouse. Spinach and cilantro are far less particular—every variety I've grown has survived winter without insulation, although growth improves and frost damage is lessened with row cover. Refer to Cold-Hardy Growing Notes (page 154) for further guidance and use the following suggestions when selecting seeds.

The name says a lot. 'January King' cabbage, 'Winter Density' lettuce, and 'Giant Winter' spinach are varieties specifically selected for cold

tolerance, whereas 'Summertime' lettuce is best saved for warmer weather. Read a variety's full description to learn about specific recommendations for winter cultivation.

Purchase from companies in zones close to yours. Varieties with proven records of success in zones like your own will be more reliable than varieties that succeed for growers in warmer regions.

Try something unfamiliar. Cress, endive, radicchio, sprouting broccoli, leeks, chickweed, arugula, and chicory are often overlooked as possible winter crops, but these are some of the most cold-tolerant plants in existence.

COLD-HARDY GROWING NOTES

Refer to the specific variety's description on the plant tag or seed packet to determine its level of cold tolerance in your hardiness zone and use the following notes to inform your fall and winter selections.

Arugula is one of the more cold-tolerant salad greens; its spiciness seems to lessen when grown in cooler temperatures. Sow fall and winter harvests later than some of the other cold-hardy vegetables because arugula seeds don't easily germinate in heat and the plants are quick to bolt. Arugula can survive unprotected in many climates, but growth is improved with insulation.

Celery takes nearly four months to mature, so seeds for fall plants need to be sown in early to midsummer, a task that can present some major challenges in warmer climates. Celery can survive winter with heavy insulation in mild climates, but it's not as hardy as some of the other recommended plants. Seedlings dry out easily, so it's best to keep them indoors, protected from the heat of the sun, until they're well established.

Rainbow chard makes a delightfully colorful addition to your winter crop selection.

Harvest entire heads or simply remove the outer stalks for continued growth.

Chard survives when grown under cover. The colorful leaves of rainbow chard become brighter and more pronounced as temperatures drop. Plants will occasionally succumb to repeated heavy frosts, but I've witnessed seemingly dead plants resprout in early spring when temperatures rise.

Chickweed—native to most US regions—germinates and thrives in winter without protection. An insulating layer will help it flourish earlier in the season but isn't always necessary. Seeds are available for purchase, but I often find chickweed sprouting on its own as a winter-garden "weed." Chickweed has a mild flavor similar to that of lettuce and can be used raw in salads or chopped and added to soups, stews, or sautés. It bears small white flowers in late winter that are also edible.

Chicory family members include chicory's close botanical relatives radicchio, endive, escarole, and frisée. Chicory and radicchio are the most cold tolerant of the bunch and can be grown without cover in many climates. Even if the outer leaves experience frost damage, the interior head remains tightly wrapped beneath a protective layer: Peel away the damaged leaves to reveal a perfectly colorful head. Escarole, endive, and frisée are slightly less cold tolerant but are worth growing for their unique colors, textures, and flavors. Insulate the greens to ensure better survival rates.

Cilantro won't germinate when temperatures remain high and is quick to bolt in heat, therefore it's sown much later in the season than other fall and winter crops. Cilantro can survive uncovered in many climates, but growth is improved when the plant is protected.

Collards, like many other brassicas, improve in flavor with some frost. Choose cold-hardy

varieties such as 'Cascade Glaze' or 'Champion' for more reliable overwintering. Insulating collards protects them from damage that can occur after repeated hard frosts.

Upland and garden cress (not to be confused with watercress) are extremely hardy salad greens that can survive uncovered in many climates. Their strong peppery flavor is best used in moderation in soups, sautés, or salads. Growth improves when the cress is insulated.

Kale does well as a winter crop. Lacinato varieties are best for overwintering, but even

Shallow-rooted greens like lettuce, spinach, mizuna, and tatsoi (shown here) can be grown in trays in a greenhouse if bed space is limited. Harvest the greens before they fully mature as the roots have limited space to grow.

EMBRACE THE BITTER

>>> Endive, radicchio (shown here), Treviso, frisée, escarole, and chicory are close botanical relatives. The exact distinctions between them aren't always clear, and they're often used similarly in culinary applications. Their crisp leaves are adorned with beautiful patterns and colors that brighten up any winter or spring meal.

Use raw leaves in a mixed greens salad, or braise, grill, or sauté halved heads to reduce the bitterness. Throwing a handful of leaves into a soup near the end of its cooking time is another great way to incorporate these plants' unique flavors.

some of the more heat-tolerant varieties like 'Red Russian' will survive when insulated. As with all overwintered crops, kales are quick to bolt in spring. Harvest the unopened flowers as an abundant broccoli substitute ('Red Russian' is my favorite for this).

Leeks can withstand a range of climatic conditions. All the varieties that I've grown have survived winters uncovered, but you'll need to select the hardiest varieties such as 'Tadorna' if growing in a cold climate. Most seed-grown leeks take four to six months to mature.

Lettuce can be succession planted through fall and early winter for continuous harvesting all winter. The most cold-hardy varieties such as 'Kweik' and 'Winter Density' prefer cover for reliable survival. Gardeners in extremely cold climates may need to double up on insulation.

Mizuna is a frilly, mild green similar to mustard greens but is most often eaten raw in salads. Mizuna won't reliably survive winters without adequate insulation. It has performed extremely well for me in the greenhouse.

Onions for fall harvest are sown in the middle of summer. Seeds sown later in the year can overwinter with insulation in some regions, but they don't always develop into large bulbs and are best eaten as scallions. Day length stimulates bulb formation, so timing is important. Plant long-day onions if you live in a northern climate with long summer days and short-day onions if you're in the South. Try growing shallots, potato onions, or walking onions (see page 68) for more reliable winter producers.

Pak choi and tatsoi are reliable winter survivors, especially when grown under row cover. Older leaves that matured in fall are more susceptible to frost damage than newer leaves that mature when temperatures remain low. Harvest

Doubling up on insulation, such as growing beets under row cover inside a greenhouse, further protects the leaves from frost damage.

Lacinato kales such as this 'Dazzling Blue' are more cold tolerant than their smooth-leaved relatives.

Leeks are reliable winter survivors that resume growth in the early spring and summer.

these older leaves in late fall and early winter. As with most overwintered greens, pak choi and tatsoi bolt quickly in spring.

Parsley varieties such as 'Mersin' or 'Darki Triple Curled' overwinter without cover in milder climates but require insulation when winter temperatures are consistently lower than 20°F (−7°C).

Root vegetables, including beets, carrots, rutabagas, and turnips, have minimal growth in cold soils, and the greens can suffer frost damage more easily than the roots. Use row cover along with a pile of dried leaves or straw (see More Uses for Row Cover, page 146) or two layers of row cover over wire or plastic hoops to keep the roots alive through winter. In regions with particularly harsh winters, harvest roots before

the soil freezes, wash and dry them, and store them in a ziplock bag in the refrigerator for up to two months.

Spinach seeds won't germinate when the soil is warm, and plants quickly bolt when temperatures start to rise in spring. Direct sow spinach at the beginning of fall and insulate it once temperatures consistently hover around freezing at night. Direct sow again in the middle of winter under row cover during a wet, warm spell if the soil isn't frozen. Direct sow a final time at the beginning of spring. Spinach will survive uncovered in many climates, especially overwintering varieties such as 'Giant Winter'. Start by following the recommended planting dates for your region, but don't be surprised if you need to narrow the suggested window to cooler months.

Netting protects these seedlings from cabbageworms.

SO YOU WANT TO GROW MORE BRASSICAS . . .

>>> I've recommended leafy brassicas such as collards and kale but haven't recommended some of the more glamorous brassicas such as broccoli, cabbage, cauliflower, and Brussels sprouts. This is because these heading versions are far more challenging to grow, especially when starting them from seed. Sow seeds too early in the spring and their growth will be stunted by frigid temperatures. Sow too late and they'll bolt before you ever get much of a head. Fall harvests have similar challenges, only there's a temperature reversal (early sowings get too hot, late sowings get too cold).

Sow seeds for starts as indicated on your planting chart and refer to Timing is Everything on the facing page.

Heading broccolis—the kind you purchase at the grocery store—don't survive frequent and repeated hard freezes and therefore are planted in early spring for a summer harvest or in late summer for a fall harvest.

Sprouting broccolis—plants that produce numerous florets rather than a single large head—are a better choice for overwintering because many varieties require a period of cold to stimulate floret production (florets form in the spring after overwintering). Insulate sprouting broccolis to protect them from damaging frosts.

Overwintering cauliflower varieties such as 'All the Year Round' and 'Prestige' are much like sprouting broccolis in that seeds are sown in the fall and heads are produced in the spring after a period of cold. Note that not all cauliflowers are suited for winter growing. Overwintering varieties need to be insulated in most climates.

Cabbage varieties have varying levels of cold tolerance, with Savoys often being the hardiest. Sow smooth-leafed varieties such as 'Golden Acre' in early spring for a summer harvest and again in mid to late summer for fall harvests. Cover mature fall cabbages on cold nights but harvest before they experience frequent, heavy frost. 'January King' and F1 hybrids such as 'Stanton' or 'Deadon' are also sown in mid to late summer, but they can more reliably last through winter when insulated with row cover.

Brussels sprouts are, in my opinion, the most difficult heading brassica. They require the longest growing season of all in that they're planted in early spring for a late-fall harvest (yes, they take that long to mature). Mature Brussels sprouts can be overwintered by insulating the plants with dried leaves or hay and tacking down the insulation with row cover.

TIMING IS EVERYTHING

When planting in spring, you're guaranteed increases in both day length and temperature as the season progresses. The opposite is true for fall plantings: Day length and temperature decrease, which means the number of days required to reach maturity can be significantly greater. A spring-planted kale can be harvested in as few as 65 days, but the same variety may take as many as 85 days when planted in fall, given the gradual reduction in light and temperature leading up to the so-called Persephone period (see Persephone: Goddess of Spring, page 160). Use the timing recommendations listed there, but don't be afraid to deviate and experiment. You'll slowly find the strategies that work for your particular situation.

In regions north of latitude 32°N and south of latitude 32°S, there is a stretch of time when winter day length drops to fewer than 10 hours of sun per day. This so-called Persephone period is longer and more pronounced in regions closer to the poles and shorter and less pronounced in regions closer to the equator. Because plants generally go dormant and nearly stop growing when they receive fewer than 10 hours of sun daily, winter and fall crops must be planted early enough that they have time to nearly mature before the Persephone period starts. Growth will be almost nonexistent, but the plants will remain alive and well. You can think of it as using the garden as living cold storage throughout winter. Refer to Appendix C to learn when to plant certain crops if they're to mature in time for winter and spring harvests.

PLAN YOUR HARVEST TIME

Seeds and starts are planted at varying times in summer and fall depending on the anticipated

Late fall sowings of lettuces, chard, kale, rutabagas, and radicchio matured slowly and were harvested the following spring.

harvest date. See Start Seeds for Four-Season Crops, page 161, for tips and more information.

Sow fall-harvest crops in mid to late summer. Get these crops in early so that they mature well before the Persephone period. These are some of the most challenging crops to grow because you're forcing cool-season seeds to germinate in the heat of summer.

Sow winter-harvest crops in fall. Planting dates are determined by counting backward from the start of your Persephone period (see Appendix C).

Sow early-spring-harvest crops in late fall and early winter. Your final round of plantings remains immature during winter but resumes growth in early spring. Overwintered crops bolt quickly and are therefore often harvested at a much smaller size. Because of this, you may or may not find these early-spring crops to be worth the effort.

This late-fall-sown spinach was harvested as baby greens throughout winter. Mature leaves were harvested for a short window in early spring before the crop bolted.

PERSEPHONE: GODDESS OF SPRING

>>> Persephone, the Greek goddess of spring, was the daughter of Zeus and Demeter. As the tale goes, Hades became infatuated with Persephone and pulled her into the underworld to become his wife. Demeter, distraught at the loss of her daughter, negotiated an agreement in which Persephone was permitted to spend eight months of the year on earth and the other four months in the underworld. Persephone's time on earth coincides with an abundant growing season (spring, summer, and fall), which ceases when she returns to be with her husband (winter).

Eliot Coleman, organic farmer and author extraordinaire, is known for his elaborate record keeping and for coining the term "Persephone" as it relates to gardening. The data he collected from his farm over the years showed that the effect of plants receiving fewer than 10 hours of sun per day was pronounced enough to give it a name. According to Coleman, "Persephone" is the period during which plants stop growing while they eagerly await the return of the goddess of spring.

START SEEDS FOR FOUR-SEASON CROPS

The soil is often hot and dry when it comes time to plant the fall and winter garden. Seeds from cold-tolerant plants are often resistant to germinating under these conditions and must be given extra attention if you're to have successful growth.

Note that not all seeds germinate in summer. Cilantro and spinach, for example, will sprout only when soil temperatures are consistently below 70°F (21°C). You may find that these types of crops are extremely challenging to grow for fall harvests and may be better suited for winter and spring harvests only.

Growers in extremely warm regions may need an indoor light setup. While not necessary for all growers, indoor grow lights can offer precise climate control at any time of year. Start the seeds indoors and transplant seedlings outdoors once temperatures have dropped.

Watering is extra important. Keep summer and fall seedlings well watered. Frequent watering, even a light spritzing, helps cool the soil and stimulate germination. In some situations, this means watering twice per day when direct sowing in hot soils that quickly dry out.

Provide shade for seedlings. When starting seeds outside, find a shady location that's protected from heavy rainfall where seeds can sprout and mature in their flats before planting out. Choose a visible location so you see them regularly and remember to water them.

Use shade cloth, a woven fabric that blocks out varying amounts of sun, when direct sowing or if shade isn't available. Growers in warmer

Fall seedlings are started during the heat of summer and therefore need to be kept cool. These fall starts were sown in August and kept in the shade under a tarp to protect them from heavy rains.

Come late winter, the greenhouse serves as a home for potted veggies, in-ground edibles, and starts that will fill the spring and summer garden.

regions need more sun blockage (50 to 75 percent shade) while growers in cooler regions can get away with less protection (30 percent shade). Shade cloth can be used to build a shaded outdoor seed-starting station (add a layer of clear plastic to protect sensitive seedlings from heavy rains) or draped over low tunnel hoops in the garden. Leave both ends open to allow better air circulation.

PLANT EXTRAS FOR WINTER

Remember that winter crops won't grow very much, if at all, unless you live in a region with minimal seasonality. For most of us, this means that the cut-and-come-again harvest method doesn't apply to winter harvests. Don't expect to continuously harvest from a single plant as you would during warmer months beccause there's little new growth until day length and

temperatures increase. Plant three to five times as much as you would in spring and summer as you can only expect to winter harvest once, maybe twice, from crops such as kale, collards, leaf lettuces, and cress.

HARDEN OFF SEEDLINGS IN FALL

Establish the seedlings in the garden during fall but don't insulate them until temps drop into the 30s and low 40s (−1° to 6°C) at night. Crops must acclimate to lower temperatures, meaning they don't require protection until it's consistently cold. Uncover or vent cold frames and low tunnels during the day when it's sunny or warm. Frequently check the interior temperature of the low tunnels to get a feel for their effectiveness. Your tunnels will likely need frequent management in fall and spring but much less in winter.

PROTECT SPRING SEED STARTS

Plants perform better when provided with real sunlight. Better yet, outdoor-grown starts don't require hardening off, thus making cold frames and greenhouses an excellent location to start seeds in spring. The internal temperature of the cold frame or greenhouse will likely be high enough to keep the seedlings warm on sunny days, but they'll need supplemental heat when it's cool and cloudy and at night.

Seed-starting mats can provide adequate heat. Place a seed-starting mat in the bottom of a cold frame or on the floor of the greenhouse with the flats placed directly on top. Commercial mats with thermostats are best because they're better at regulating fluctuations in temperature. If the cold frame is small enough, it'll trap the heat from the mat and likely keep the seeds warm enough overnight. When using a mat with a thermostat, you can drape a layer of row cover directly over the seedlings to provide even more insulation (do not do this with unregulated-temperature mats; that would trap far too much heat).

Use a foam insulator on a grow bench. Placing seedlings directly on the ground in the greenhouse is best because the ground will provide additional heat. If you must use a bench, first put down a piece of insulating foam to help trap the heat from the seed-starting mat.

Moving plants inside and outside is risky. Seedlings can be moved indoors at night and set back out in the cold frame or greenhouse the following morning if you don't want to purchase a seed-starting mat. However, this can become extremely labor intensive depending on the number of seedlings you've started—and just one night of forgetfulness can be a costly mistake.

Heat released from seed-starting mats is adequate to heat the interior of a small cold frame but will have almost no impact in a large greenhouse. Place one or two layers of row cover or even a cold frame on top of the mats inside the greenhouse to act as an insulator. Remove covers on sunny days and replace them when the temperature inside the greenhouse starts to drop as the sun sets.

6

larger-scale

STORAGE CROPS

Growing storage crops like dried beans, potatoes, sweet potatoes, and winter squash allows you to enjoy a garden's bounty long after the fruits of summer have come and gone. Plants and seeds placed in late spring and early summer sprawl into remote corners of the garden. Generously allot space for these plants and be rewarded with unique and delightful storage crop varieties unavailable in most grocery stores.

Carbohydrate-rich vegetables take up a great deal of garden space, especially if you expect to harvest an entire year's supply.

the plan

- **EXPAND YOUR GARDEN BY 600 SQUARE FEET.** Sweet potatoes and winter squash need space to sprawl, and corn requires a minimum of 20 plants, so this will be your largest expansion yet. If possible, position these beds with plenty of room for vining plants to grow away from the beds instead of taking up valuable growing space within them.

- **SELECT REGIONALLY APPROPRIATE VARIETIES** and plant at the right time. Pay close attention to the length of your growing season and the chosen variety's days to maturity. Growers with shorter seasons may have trouble growing larger winter squash and pumpkins compared to smaller varieties that typically mature quicker. Sweet potatoes aren't easily grown in the coldest of climates. Growers in any climate can grow beans for drying, especially quick-to-mature bush varieties. Pole beans take slightly longer but flourish in a wide range of climates.

- **CONSIDER LONG-TERM STORAGE NEEDS.** Except for corn, starchy vegetables can be stored for months at or below room temperature. To prolong their shelf life, store them in a root cellar, cool basement, or insulated garage (the interior temperature should ideally remain below 50°F/10°C but never drop below freezing). Freezing cooked starchy vegetables is also an option. Beans will keep for a year or longer at room temperature if they've been properly dried.

BEFORE YOU DIG IN

Starchy vegetables require a large investment of time, energy, and space. You therefore want to focus on growing the vegetables that you'll use and enjoy while also recognizing that some will be easier to grow than others. Start by selecting a wide range of vegetables and varieties to see how they perform. A failed attempt one year may be corrected by choosing a different variety the following year. Over time, you'll discover the vegetables and varieties that are most worth your time.

Beans and starchy vegetables need a lot of sun. Some winter squash, beans, and potato varieties can be successfully grown in as few as six hours of sun per day, but production will improve with increased sun exposure. For corn and sweet potatoes, the hotter and sunnier, the better.

Leave space for sprawl. Vining winter squash and sweet potatoes can spread as far as 20 feet. Directing vines away from planting areas optimizes space utilization throughout the growing season, so orient the new beds so that there's space for sprawl in at least one direction.

Think about using trellises, too. Pole beans are always grown along trellises, but winter squash can be as well. There are benefits and limitations to each of these growing methods (see Prevent Vine Borer Damage, page 173), but saving space is the primary reason to trellis pumpkins and winter squash.

Focus on long-term soil nutrition. As always, build the beds by layering in lots of organic matter—dried leaves, sticks, wood

Some winter squash varieties can sprawl up to 20 feet. Train them to grow outside of beds to keep them from taking up valuable gardening space.

EXPLORING ROOT CELLARS

>>> Root cellars—underground food storage rooms—have been used for thousands of years. Dig into the ground deep enough (about 10 feet, sometimes as little as 5) and you'll find a constant temperature of around 40°F (4°C). A root cellar, traditionally a small room with stone or cement walls and a dirt floor, acts as a refrigerator in summer and prevents food from freezing in winter.

The exact depth of a root cellar and its location depends on climate, soil type, and water table depth. Cellars in colder regions need to be placed below the freeze line and should be positioned on a south-facing slope to capture more heat. Cellars in warmer regions are better positioned on north-facing slopes and may not need to be sited as deep. Search online for more specific root cellar guidance such as determining a cellar's depth, construction recommendations, and the effects of humidity, temperature, light, and ventilation on food storage.

Spaghetti squash were trellised alongside 'Cherokee Trail of Tears' black beans.

chips, mulch, logs, and compost. Heavy feeders like squash and corn require loads of nutrition. Starting the beds correctly will ensure production for years to come.

Don't forget to rotate crops. Starchy vegetables demand a great deal of nutrition from the soil. Beans, on the other hand, can improve the soil by drawing in nitrogen via a symbiotic relationship with mycorrhizal fungi. Alternating between light and heavy feeders and top-dressing with compost improves soil health and production. If rotation isn't possible, plant a winter cover crop such as daikon radishes or field peas after a heavy-feeding plant is spent.

LEARN ABOUT COMPANION PLANTING

Certain crops are thought to benefit from being planted together. They might share complementary nutrient utilization or growth habits, or one might help the other repel insects. Likewise, some plants are thought to perform poorly when grown together. The best-known example of successful companion planting is the Three Sisters, a system developed by Indigenous peoples of interplanting corn, beans, and squash.

ARK OF TASTE

>>> The Ark of Taste was started in 1996 to document and catalog culturally significant foods—vegetables, livestock breeds, spices, grains, beverages, preserves, breads—that are at risk of being lost whether it be from biological, commercial, or cultural pressures. Their collection contains more than 3,500 rare and important products from 350 countries. Inclusion in the Ark of Taste requires that a food be sustainably grown, have a unique and/or superior flavor, and represent local or regional traditions. Look for the Ark of Taste symbol when selecting seeds if you're interested in supporting this important work.

Ark of Taste

According to the *Old Farmer's Almanac*, many companion plant suggestions are based on lore. In most instances, a plant can benefit from being planted with *any* other plant; you don't have to overthink it. In general, avoid planting members of the same family together (in other words, don't plant dill with carrots), with the exception of brassicas, which are best planted together so that they can be easily covered with insect netting or row cover. Also avoid plantings that will compete due to similar growth habits, such as tomatoes and sunflowers, which need lots of vertical space.

SOME SUGGESTED COMPANION PLANTINGS

Here are a few common combinations, but you can rarely go wrong with any interplanting that enhances diversity.

asparagus

- Calendula

basil

- Lettuce
- Peppers
- Tomatoes

beans

- Beets
- Corn
- Nasturtiums
- Squash

beets

- Alliums
- Brassicas
- Bush beans
- Lettuce

broccoli

- Other brassicas

cabbage

- Nasturtiums

carrots

- Cabbage
- Chives
- Leeks
- Lettuce
- Onions
- Peas
- Radishes
- Tomatoes

corn

- Beans, pole
- Cucumbers
- Dill
- Marigolds
- Sunflowers
- Melons
- Peas
- Squash

cucumbers

- Beans
- Dill
- Lettuce
- Nasturtiums
- Radishes
- Sunflowers

eggplant

- Carrots
- Onions
- Peppers

lettuce

- Basil
- Beets
- Cabbage
- Carrots
- Chives
- Onions
- Radishes
- Scallions
- Spinach

okra

- Basil
- Cucumbers
- Lettuce
- Radishes

onions

- Beets
- Cabbage
- Carrots
- Chard
- Lettuce
- Tomatoes

continued on next page

peas

- Alyssum
- Beans
- Carrots
- Chives
- Corn
- Cucumbers
- Radishes
- Turnips

peppers

- Basil
- Carrots
- Onions
- Tomatoes

potatoes

- Basil
- Beans
- Calendula
- Peas
- Squash

radishes

- Lettuce
- Nasturtiums
- Peas

spinach

- Beans
- Brassicas
- Cilantro
- Peas

tomatoes

- Asparagus
- Basil
- Calendula
- Carrots
- Celery
- Chives
- Cucumbers
- Nasturtiums
- Onions
- Parsley
- Peppers

summer squash and zucchini

- Nasturtiums
- Zinnias

winter squash and pumpkins

- Beans, pole
- Calendula
- Corn
- Marigolds
- Nasturtiums

EXPLORE HEIRLOOM VARIETIES

Gardeners go gaga over heirloom tomatoes, but the assortment of heirloom beans, winter squash, and pumpkins may be even more impressive. We equate pumpkins to jack-o'-lanterns, winter squash to butternut, and dried beans to pinto beans, but this narrow perspective fails to represent the breadth of flavors, sizes, colors, and stories embodied by many historically significant foods. Storage foods were vital before canning, refrigeration, and freezing were available to preserve summer's bounty.

Pumpkins, squashes, potatoes, and beans along with grains, ferments, fats, and animal products have sustained civilizations long before chain grocery stores carried specialty items like hearts of palm or coconut aminos. These stories are now locked within the genetic makeup of a wide range of lesser-known varietals.

Some of my favorite seed companies for heirloom beans, corn, potatoes, winter squash, and pumpkins are Uprising Seeds, Southern Exposure Seed Exchange, Fruition Seeds, and Seed Savers Exchange. They offer varieties dating back as far as the early 1600s that can be traced to varying Indigenous peoples across the globe.

'Cherokee Trail of Tears' (a.k.a. 'Cherokee Black') beans, for example, were carried to Oklahoma by the Cherokee people in the early to mid 1800s when they were driven from their land in the Smoky Mountains. 'Green-Striped Cushaw' squash is believed to have originated in Mesopotamia as early as 7000 BCE.

There's no reason to avoid modern varieties such as 'Nodak' pintos, which were specifically bred to mature quicker in North Dakota's short growing season, but I think you'll enjoy learning a bit more about the historical significance of storage foods and the people who relied on them for survival.

GROW WINTER SQUASH AND PUMPKINS

Winter squash and pumpkins are warm-season plants. Seeds or starts should therefore be placed in the ground after all threat of frost has passed. Fruits grown from starts will mature earlier in the season, which can be problematic in warmer climates when it comes to storage: The fruits will be ripe in the middle of summer, when cool temperatures aren't readily available.

This won't be an issue for those with a root cellar or cool basement, but my garage (where most of my storage foods are kept) is too warm to store squash in summer. Early squash and pumpkin harvests will often rot in the warmth. If you don't have early-season cold storage, plant just a few starts early on but save the bulk of the seeding until after the last frost date.

SELECT YOUR VARIETIES

Winter squash and pumpkins come in all shapes, sizes, colors, and flavors. Read the seed descriptions to gain a better understanding of each selected variety, but also check with local growers to learn which varieties work best in your region. To help you get started with variety selection, consider the following attributes.

Size. Sweet dumplings and honeynut squash are small, single-serving squash. Candy roasters and 'Blue Hubbard' each weigh 10 to 20 pounds and are large enough to feed a small army. I prefer small to medium-size squash and pumpkins because they're easier to cook and don't require that you eat squash every day for an entire week. Other growers relish the sport of producing impressive monstrosities from their backyard gardens.

Growth habit. Most winter squash and pumpkins are vining, but a few grow as bush or semi-bush varieties: Look for 'Emerald Bush' buttercup, 'Burpees Butterbush', 'Table Gold' acorn, sweet dumplings, 'Zeppelin Delicata', and 'Table Queen Bush'. Bush varieties are approximately 3 feet in diameter; semi-bush ones vine up to 4 or 5 feet.

Days to maturity. Winter squash and pumpkins require 85 to 125 days to reach maturity depending on the variety. Smaller varieties generally take less time to mature than larger cultivars. Choose varieties with different days to maturation to extend the harvest window. Growers with extremely short seasons should start seedlings indoors or select varieties with fewer days to maturity.

The vining habit of squash can make an excellent accent for beds filled with mixed perennials and edibles. Choose varieties that can become an integral part of your edible landscaping if possible.

When the vines and leaves begin to die, it's time to harvest winter squash.

Culinary use. People are most familiar with the texture and flavor of acorn and butternut squash, but squash and pumpkins are available in a wide range of sweetness levels and textures. Sweet, dry squashes and pumpkins such as kabocha, 'Red Kuri', 'Black Futsu', crown pumpkins, and 'Robin's Koginut' can easily be combined with stocks and broths to make creamy soups or baked into firm casseroles or pies, sometimes without the addition of a sweetener. The peels of many squashes are edible, so don't let a bumpy skin deter you from a particular variety. For example, bumpy 'Black Futsu' is nearly impossible to peel but can be sliced, diced, and roasted, peel and all.

Storage life. Recently harvested squash and pumpkins can be stored for as little time as four weeks or as long as an entire year, variety dependent. Varieties with a short storage life need to be cooked and frozen or canned if you're

SQUASH, PUMPKIN, OR GOURD?

>>> Winter squash, pumpkins, summer squash, and most gourds fall within one of four groups: *Cucurbita moschata*, *C. maxima*, *C. argyrosperma*, or *C. pepo*. Winter squash and pumpkins are usually *C. moschata*, *C. argyrosperma*, or *C. maxima*; summer squash and gourds are typically *C. pepo*. There are, of course, exceptions. These distinctions become important when it comes to saving seed (page 186).

Summer squash such as yellow crooknecks, zucchinis, and pattypans are bred to be picked and consumed when they're small and tender. Left to mature for a few months, they grow larger and develop a hard outer skin, and the seeds mature into something resembling pumpkin seeds—they become more like a winter squash, pumpkin, or gourd.

Winter squash and pumpkins can be consumed when they're young and tender, but they're tastiest once they've developed a hard outer skin and the vines begin to die back. Gourds aren't generally consumed at any stage due to subpar texture and a bitter flavor. They're instead used for functional pieces like birdhouses, storage vessels, ladles, and bowls or for decorative purposes.

to keep them on hand year-round. Many of the larger pumpkins and squash have longer storage life and can be kept in a cool garage or root cellar for year-round consumption.

Insect and disease resistance. Squash and pumpkins are prone to vine borer infestations and powdery mildew (see Prevent Vine Borer Damage at right). Some varieties are resistant, but there's never a guarantee. Try borer-resistant varieties such as butternuts and cheese pumpkins if you're unable to grow other varieties due to pest pressures.

Pollination and seed saving. Squash and pumpkin seeds are among the more challenging types to save because they readily cross with other squash, zucchini, and pumpkin varieties to yield Franken-gourds. You rarely get a fruit that looks like the parent when saving seed from squash plants. See Introduction to Seed Saving, page 186, to learn how to successfully save seeds from squash.

Viner borer moths lay their eggs on squash plants, and the larva bore their way into squash stems.

PREVENT VINE BORER DAMAGE

Winter squash can be grown on a trellis to save space (although I wouldn't do this with extremely large varieties because of fruit weight) and to encourage evenly ripened fruits, but there's a small risk associated with trellising from a pest called the squash vine borer.

Trellised squash vines have just a single root system where the squash grows out of the ground. In contrast, squashes grown along the ground have multiple root systems—stems that touch the soil produce secondary roots from nodes along the length of the vine. Vine borer moths lay eggs on squash plants, and the larva bore their way into the vines—most often on the portion of the stem directly above the root. The plant's vascular system is destroyed as the larva feed.

Because a trellised squash has just a single root system, a vine borer infestation can be fatal if that system gets cut off from the aboveground biomass. But only in rare instances will vine borer infestations be so prolific as to destroy the primary *and* multiple secondary root systems on a squash grown along the ground. These secondary root systems can save a squash's life.

Signs of squash borers include small holes in the vines near the base of the plant or base of the fruits, with sawdustlike residue collected around the hole. Gardeners don't often notice the damage until it's substantial. Preventing borer infestations can be challenging. Some experts recommend wrapping vines with aluminum foil, which seems like a giant pain, so I don't do it, but I rarely trellis my squash, either.

Vine borers will sometimes bore into the fruits or the stems of fruits. Hard-stemmed varieties such as butternuts, honeynuts, koginuts, or crown pumpkins are more resistant. Growing squash along the ground and planting borer-resistant varieties can make or break your squash success in any given year.

KNOW WHEN TO HARVEST

Winter squash and pumpkins should be left to mature until the vines begin to turn brown and die. Use pruning shears to clip the squash from the vine, leaving a 1-inch stem on the fruit. Cure the squash at around 80°F (27°C) for 10 days to help harden the skin and to seal any injuries. Transfer the squash to a cool, dark place (around 50°F/10°C) for long-term storage. Squash should be stored in a single layer with good ventilation, such as on a wire or wooden rack.

A few of my favorite squash varieties are 'Robin's Koginut', 'Red Kuri', 'Black Futsu', and '898'.

SOME GOOD CHOICES

>>> Over the years, I've experimented with several different squash and pumpkin varieties. I now have a set of tried-and-true varieties that I plant every year along with something new and different. These are all small to medium-size dry, sweet varieties, most with edible peels.

- 'Robin's Koginut' squash (from Row 7 Seed*)
- '898' squash (from Row 7 Seed*)
- 'Black Futsu' pumpkin
- 'Red Kuri' squash
- 'Black Forest' kabocha

*Note: The varieties from Row 7 Seed Company are hybrids specifically bred for superior flavor. Check out Row 7 Seed's website to learn more about a company that stemmed from a unique partnership between chefs and plant breeders. They have a limited selection of incredibly delicious vegetables.

For Fun, Not Food: Grow a Birdhouse

Gourds have growing requirements similar to those of squash and pumpkins. Birdhouse gourds harden as they dry and are grown for utilitarian purposes, not eating. A single plant can produce 10 or more fruits.

The pale green gourds become speckled with fungus as they start to harden. The vines brown and wither as the gourds transform into hard, brown orbs. Cure the mature gourds in a warm, dry spot for three months. Before using them, scrub well with soapy water and a stainless steel scouring pad.

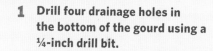

1 Drill four drainage holes in the bottom of the gourd using a ¼-inch drill bit.

2 Drill two ¼-inch holes at the top of the gourd and insert a wire or string for hanging.

3 Use a hole saw to cut an entrance hole* in the largest part of the gourd, then remove all the seeds.

4 Paint or decorate the birdhouse if you wish. Hang it in early spring, 6 to 10 feet from the ground.

*Different-size entrance holes attract different bird species. Smaller gourds (4 to 5 inches) need a 1¼-inch hole to attract small birds such as wrens, chickadees, nuthatches, and titmice. Bluebirds, swallows, and small woodpeckers prefer a 5-inch gourd with a 1½-inch entrance hole. The largest gourds (7 inches and larger) can accommodate a 2- to 2½-inch entrance for larger birds like flickers, purple martins, flycatchers, and red-headed woodpeckers.

SWEET POTATOES LIKE IT HOT

Sweet potatoes are a member of the Morning Glory family. Expect plentiful vines and even a few trumpet-shaped flowers as the plants mature. Sweet potatoes prefer consistent soil temperatures above 80°F (27°C), good drainage, and moderate nutrient availability. Ideal growing conditions can be slightly difficult to achieve unless you live in a region that traditionally grows them (the southeastern United States).

Choose varieties such as 'Beauregard', 'Centennial', 'Burgundy', or 'Georgia Jet' if you have a short growing season. Growers in warmer regions can grow whichever varieties they'd like.

Grow your own slips. Portions of sweet potato vines are clipped from sprouted potatoes and placed in water to root. These cuttings, called slips, give rise to new plants. Slips are planted 4 to 6 inches deep with the lowest leaves remaining above the soil line. You can occasionally find starts grown in six-packs, but slips are far more common. See Growing Sweet Potato Slips, page 178.

Sweet potato vines can sprawl up to 20 feet. We planted these at the edge of the bed where the vines could sprawl into a large mulched area without taking over another crop.

Prepare the bed. Sweet potato vines can extend up to 20 feet in any direction, so plant them with plenty of room to sprawl. You can keep the vines trimmed to 3 feet in length, but I find this to be tedious.

Sweet potatoes prefer sandy or loose soils; they won't grow well in compacted clay soils. If you have heavy soil, loosen it with a broadfork (see page 63) or digging spade prior to planting. Planting in loose, organic matter such as decomposed wood chips and dried leaves can have the same impact as over-fertilizing, which leads to the next point: Don't over-fertilize the soil. Sweet potatoes were traditionally grown in low-nutrient soils, but production improves with moderate nutrient content. Amend the beds with a small amount of well-decomposed compost prior to planting but don't over-fertilize, especially with nitrogen. Too many nutrients can result in irregular, elongated roots.

Plant at the right time. Sweet potato slips are best planted a month after the last frost date to ensure that air and soil temperatures are high enough to support growth. Overnight temperatures should consistently be at 68°F (20°C) for best results. Earlier plantings should be grown under row cover.

Warm the soil if you have a short growing season. Cover beds with black plastic or low tunnels a month before the last frost date to increase soil temperatures and to prepare for earlier plantings.

Don't overwater. Water the slips well for two to three weeks after planting to encourage root growth. Plants require moderate water thereafter. Too much water can cause the potatoes to grow abnormally large, have dull flavor, or crack.

Sweet potatoes grow long and slender when cultivated in extremely nutrient-rich soils. The potatoes are still perfectly good to eat, but they're harder to wash and culinary use is limited. Don't add compost before planting if you have highly nutritious soils.

The largest potatoes develop where the slip was planted. Smaller potatoes are borne along the vines where secondary roots grow into the ground.

CRITTERS LOVE CARBS

>>> Carbohydrate-rich vegetables take longer to mature than many of the other vegetables you've grown thus far. Pests therefore have a longer window of opportunity to locate and nibble on the crops. Furthermore, humans aren't the only organisms that rely on carbs. Pest pressures could become more diverse in that your garden may attract raccoons, squirrels, voles, crows, and deer for the first time ever.

Adjust your expectations or practices if you find that pest pressures become so troublesome that it's difficult to grow certain foods. For example, I've never been an advocate for numerous raised beds because the investment of time and materials doesn't easily pay for itself. But my vole pressures are so substantial some years that I find it difficult to grow roots and tubers such as potatoes, sweet potatoes, beets, carrots, and turnips. I'm now transitioning a portion of my garden over to raised beds lined with hardware cloth for use specifically with roots and tubers (see image on page 72). The hardware cloth will prevent the voles from burrowing into my beds and destroying my crops.

Other strategies for guarding crops against pests include planting early, using succession plantings, covering the beds with insect netting, and building physical barriers such as fences. But sometimes the best option is to move on and instead focus on growing crops with a proven record of success.

Growing Sweet Potato Slips

Growing your own slips saves money and gives you more control over the planting date. Start the potatoes indoors, three months before you plant out. It's best to grow the slips from a known variety that has an appropriate number of days to maturity for your region.

1 Gently pierce a sweet potato with three or four skewers so it can rest on a glass or jar with about half of the potato suspended in water. Place the potato pointed side down in the water.

2 Place the potato in a sunny window. Change the water once a week and top up as needed. The top will eventually grow sprouts. Each potato should develop 5 to 10 slips.

3 Clip off the sprouts once they're 12 inches long and place the clippings in water to grow roots. The slips are ready to plant when the roots are 2 inches long.

HOW TO HARVEST AND STORE SWEET POTATOES

Harvest potatoes after 90 to 125 days of growth. Sweet potato vines die after a frost, and the potatoes will quickly begin to degrade, so harvesting before a frost is best. If you aren't able to get to them before a killing frost, harvest no more than one or two days after.

Use a digging spade to search for smaller potatoes that have developed along the length of the vines. Periodically lifting the vines during the growing season can prevent their rooting along the soil, thus reserving all the plant's energy for the main potatoes. However, this practice becomes impractical when growing a large crop. The smaller secondary potatoes can be harvested for eating or left in place to decompose and feed the soil.

Cure sweet potatoes before storage. Rinse the potatoes after harvest and spread them out to dry in the sun for a couple of days. Set damaged potatoes aside to be eaten first. Place the remaining potatoes in shallow cardboard boxes and store in a cool, dark place such as a garage, basement, or root cellar. Sweet potatoes can be stored for up to six months under ideal conditions.

GROW DRIED BEANS

I used to think that producing a substantial quantity of dried beans would demand too much space and effort. Little did I know that beans are some of the easiest plants to grow and that harvesting dried beans is simpler than you might expect.

Dried beans have the same two growth habits—bush or pole—as beans eaten fresh (green beans). Bush beans have a bushy habit and grow unsupported, while a pole bean naturally tendrils its way up a trellis. Beans eaten fresh are picked when the pods are young and tender. Left to mature, the seeds and pods of any variety will become hard and dry. Most bean varieties are categorized as either a fresh bean or a dry bean based on flavor and texture. Some varieties such as 'Tiger's Eye' bush dry beans are considered dual purpose: Their pods are tender enough to be eaten when young but are generally left to mature as dried beans.

Pole varieties require less space, can produce heavy yields, and have better pest resistance.

Cattle panels bent into arbors are my first choice for pole bean trellises. Bend the panels over walkways to save space and to beautify your garden.

They are the best choice for growers in wet climates as their vertical growth habit allows for good air circulation, thereby resulting in fewer issues with fungus. The beans mature at different rates, which means you're not harvesting a massive yield at one time. Some varieties can grow as tall as 12 feet; design your trellises accordingly.

Bush varieties mature quicker, require more space, and have more uniform ripeness. Follow spacing guidelines for the selected variety, because overcrowding will impede airflow and encourage the growth of fungus.

HOW TO HARVEST

Most sources state that pods should be brown and dry before harvesting. However, I've found that waiting that long increases susceptibility to mold and premature sprouting within the pods. Growers in wet climates should start harvesting when the pods feel leathery and are easily peeled open to reveal mature beans.

Bush varieties are often harvested in their entirety when the leaves start to die back and the pods begin to dry. Clip off the plants directly above the roots and hang them upside down indoors to finish drying. Or remove the beans from their pods immediately and spread them out to dry.

Beans are easiest to remove by hand when the pods are still soft and leathery; the beans pop right out of their shells, making it easy to separate the pods from the beans. Alternatively, transfer the pods to a paper bag or basket and allow them to dry completely. Once dry, crack open each pod to remove the beans. Hand-harvesting at this stage can result in a few dried pods being accidentally dropped into the beans.

For large harvests, it may be necessary to thresh and winnow the beans. Thresh the beans by placing crispy, dried pods into a pillowcase or

'Nodak' pintos were harvested when the pods were soft and leathery. The pods peel right open at this stage, and the beans are easy to remove.

bag. You can crush the pods by hand, by stepping on the bag, or by banging the bag repeatedly on the ground. I've even seen growers wrap extremely large harvests in tarps and drive a tractor over the beans. Once the pods are shattered, the beans are separated out by winnowing. Traditionally, winnowing was accomplished by placing the beans in a bowl and pouring them into another vessel on a breezy day. They can also be tossed up and down in a basket. The heavy beans fall straight back down into the basket as the light husks are blown away.

Use an oscillating fan to make the process even simpler. Start with the lowest setting and bump it up, if needed. Pour the beans from one vessel into another directly in front of the fan. The pods will blow away and the beans will fall into the new vessel.

HOW TO STORE

After collection, spread the beans in a single layer on cookie sheets or screens to let them dry for another few weeks. Beans that were hand-harvested when the pods were leathery will need additional time. Large harvests require a great deal of space for drying, which is why I dry my beans in batches at 120°F (49°C) in a dehydrator. The time required to dehydrate beans depends on the size of the beans, how much moisture was in the bean to begin with, and the relative humidity. Start by drying them for 12 hours and increase the time as needed.

Once the beans are completely dry, place them in the freezer for at least 48 hours or in the oven or dehydrator at 160°F (71°C) for 30 minutes to kill bean weevil eggs. The beans can now be safely stored in sealed containers at room temperature for a year.

GIVE CORN A TRY

Of all the foods I've grown, corn is the most challenging. The stalks grow beautifully but then raccoons, deer, and squirrels enjoy nearly the entire harvest. One year, a bear knocked over every cornstalk in my garden. Animal pressures are substantial enough that I pause each time I sow corn. Many of the organic farmers in my region have stopped growing it altogether.

However, one person's failure doesn't mean another grower won't succeed. The pressures at my site may be completely different from your own. For example, growing in an urban environment, having an outside dog, or installing tall fencing around the beds may deter most four-legged critters. You should give corn a try, knowing that you're not alone if it turns out to be a crop with minimal success.

Be thoughtful when selecting seed corn. Not all varieties are well suited for eating fresh. Some are dried and used for popcorn or ground into either fine or coarse flour for baking or making grits. Select a variety with a purpose that aligns with your use. Varieties shorter than 6 feet tall with fewer days to maturation are safer bets for the first-time corn grower.

Be thoughtful when selecting seed corn. Not all varieties are well suited for eating fresh.

UNDERSTAND CORN POLLINATION

Male flowers atop each stalk provide pollen to the female flowers (immature ears of corn) below. Every tassel that extends from a female flower transforms into a single kernel if that tassel receives pollen. Corn is wind pollinated, so much of the bountiful pollen produced by male flowers blows away and never reaches the tassels below. To maximize pollination, plant at least four rows of corn with at least five plants in each row. This helps trap some of the pollen among the plants, ensuring that adequate pollen reaches the tassels. Increasing the quantity of plants further improves pollination, so having more than 20 plants is even better.

Some growers will plant just 10 plants knowing that they'll have to pollinate by hand. To hand-pollinate, secure a paper bag over an entire male flower and shake gently yet thoroughly to collect the pollen. Next, place the bag over all the tassels on a single female flower (preferably on a different plant for cross-pollination) and shake again. Repeat until the pollen from every male flower was collected and deposited on every female flower. Do this every day for a week starting when pollen grains are visible on the male flowers. If unsure of the pollen status, lightly shake the plant or blow on the male flower to see if a small puff of pollen is released.

Short varieties such as 'Painted Mountain' mature quicker and are often a better choice for the first-time corn grower. Sow clover seeds between rows when corn reaches 6 inches high to supply nitrogen to the young plants.

ELUDE EARWORMS

Earworms make their way into ears of corn and devour maturing kernels. Generally speaking, just one or two earworms are found in each ear, as they also enjoy eating other earworms—only the strongest survive. Sowing corn early helps prevent earworm infestations because earworms appear later in the season. Unfortunately, corn has sensitive roots and is one of the few plants (another is beans) considered unsuitable for starting in flats, yet it can't be sown in the ground until the soil warms and all chance of frost has passed.

To get around this dilemma, cover beds with black plastic or row cover to warm soils as early as possible in preparation for direct sowing. Better yet, sow corn seeds beneath a large cold frame that can be removed once the stalks are well established. Select varieties that are quicker to mature and more likely to be harvested before earworm season begins.

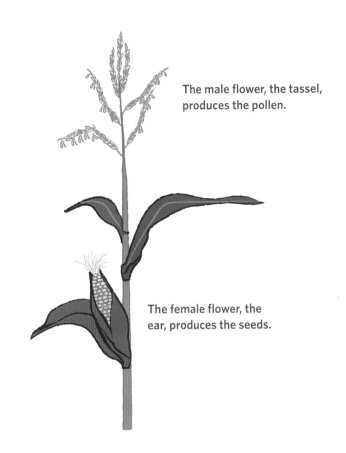

The male flower, the tassel, produces the pollen.

The female flower, the ear, produces the seeds.

KEEPING CROWS AWAY

>>> Crows eat blueberries and other ripe fruits, and have been known to destroy seedlings, especially those from larger seeds like beans, peas, and corn. Crows identify emerging seedlings, then pull the seedlings from the ground to devour what's left of the seed.

Starting seeds in flats and transplanting starts into the garden prevents crow damage, but transplanting isn't always desirable. In those instances, sow the corn, peas, or bean seeds and cover the rows with insect netting. Leave the netting in place until the plants are at least 5 inches tall, preferably longer. Remove the netting when the seedlings are at least 10 inches tall.

Use insect netting over bean and corn seedlings to protect them from being eaten by crows and other animals. It can be removed once the plants are at least 10 inches tall.

Some gardeners recommend regularly spraying the tops of maturing ears with mineral oil, vegetable oil, or the organic pesticide Bt (*Bacillus thuringiensis*; see page 145) to prevent earworms. Spray when the silks start to turn brown and dry out and continue once per week until harvested. I find pest-treatment regimens that involve weekly applications difficult to maintain, and I avoid the use of even organic pesticides, so I opt for early plantings instead.

A nearly perfect corn crop was devoured overnight by some type of critter, likely a raccoon.

THE TRANSPLANT EXPERIMENT

>>> Most gardeners say you should never plant corn and beans in pots and transplant them into the garden. However, after 15 years of growing starts from seed, I've learned how to be especially attentive with these sensitive root plants; nonetheless, this isn't a task for beginners. As you master seed starting and want to try transplanting corn, sow seeds in 2-inch pots three to four weeks before your last frost date. If space allows, sow in 4-inch pots five to six weeks early. (There will be slight variations among varieties, so use these timing guidelines as a starting point and adjust accordingly.) The larger pots allow the roots to better develop before being transplanted so they are less susceptible to damage. Sow beans in 50-cell seed-starting trays two to three weeks before transplanting.

Use as much care as you can when removing the seedlings from their pots. For corn, bury the rootball below the soil line. Corn, like tomatoes, can benefit from being deeply planted. Stunted plants or those that don't produce were too heavily disturbed during transplant. If space allows, try a side-by-side comparison using the same variety—plant half of the crop by direct sowing and the other half as transplants. This way you'll know for sure whether your transplanting efforts were a success.

KNOW WHEN TO HARVEST

It's not always easy to know when to harvest corn. Calculate the expected harvest date based on the chosen variety's days to maturity. Mark your calendar or set a reminder on your phone. Check the corn frequently around this date by squeezing the developing ears to see if they feel full and mature.

Sweet corn. Harvest corn on the cob when the silks are brown but the husks are still green. If you think it's ready, shuck a single ear and pierce a kernel to see if it's hit "milk stage." Clear liquid coming from the kernel indicates immature corn, white (milk stage) indicates that it's ready to be eaten fresh, and lack of liquid indicates that the corn is overripe.

Other varieties. Harvest popcorn and flour corn when the outer husks are dry. Pull off the ears and peel away—but don't remove—the papery husks. Tie the husks together in small bundles and hang the bundles to dry in a sheltered area. Once the kernels are fully dry, use a corn sheller to strip them from the cob. (Handheld corn shellers for flour corn and popcorn are readily available online.) Use a grain mill or a high-powered blender to grind dried corn into flour.

LET'S TALK ABOUT NIXTAMALIZATION

>>> Like all grains, corn contains antinutrients (compounds that bind with nutrients, making them unavailable for uptake) and gut-damaging compounds such as lectins. Soaking, sprouting, or fermenting grains and flours breaks down harmful compounds so our bodies can acquire the most nutrition from these foods. Nixtamalization does the same, but with this processing method, corn is soaked and boiled in lime water prior to grinding. Wood ash was traditionally used for this process, but we now use pickling lime, a common canning ingredient.

Corn flour that has been processed in this way is known as masa harina ("flour dough") and is a key ingredient in Mexican and Central and South American dishes such as tortillas, tamales, and sopes. Without nixtamalization, corn flour won't easily bind together to form a dough and is therefore better suited for use in grits or corn bread. Check out videos on how to make masa on YouTube if you're interested in trying out this useful technique.

INTRODUCTION TO SEED SAVING

Saving seed seems easy enough, but it requires some knowledge of pollination. A variety can potentially lose desirable traits when pollinated by a completely different variety. Ensuring genetic integrity from year to year requires that each variety be pollinated by another plant of the exact same variety.

Select open-pollinated varieties. When pollinated by the same variety, open-pollinated varieties remain true to type. For example, a 'Cherokee Purple' tomato (an open-pollinated variety) pollinated by another 'Cherokee Purple' tomato will yield 'Cherokee Purple' seeds. That same tomato pollinated by an 'Indigo Apple' tomato (also an open-pollinated variety) will yield seeds with unknown outcomes.

Don't select hybrids. People often confuse hybridization with genetic modification, but the two processes are entirely different. Hybridization describes the intentional crossing of two different varieties to form a third variety. Hybrids have "F1" or "F2" at the end of their name signifying how many crosses were made to create that particular variety. Genetic modification describes the actual insertion of genes, most often to grow a crop that's pesticide resistant or produces its own pesticide.

Hybridization has been occurring for thousands of years as people selectively bred varieties with highly valuable traits to produce a cross. Hybrid varieties offer improved production, unique flavor, disease resistance, or interesting shapes or coloring, but there's no guarantee that these same traits will be present in plants grown from hybrid seed. Additional crossing will result in an unknown outcome. Avoid hybrids when seed saving to preserve parental characteristics.

The distance between varieties matters. Wind-pollinated plants such as corn can spread pollen up to a mile. Because corn readily accepts pollen from other varieties, it's recommended that each corn variety be planted at least ¼ mile from another. This means your own corn seed can become "contaminated" by your neighbor's and thus may not be the best choice for seed saving. Plants such as beans rarely cross-pollinate, instead they self-pollinate. Saving seed from beans requires just 20 feet of space between varieties (see Planting Distance for Seed Saving, page 191).

Flowering time matters. Providing plants with the necessary spacing distance for seed

Start by saving seeds from some of the easiest plants including tomatillos, beans, dill, tomatoes, and many flowers.

saving isn't always possible, especially in small gardens. In this case, pay attention to flowering time. For example, early-spring kales bolt at similar times. Rather than letting all varieties flower simultaneously, select just one variety to flower and remove the others. When possible, stagger the plantings through the season to isolate flowering windows.

Some seeds are easier to save than others. Members of the Cucurbit family— squash, cucumbers, melons—readily cross-pollinate. Melons can't cross with a squash and a squash can't cross with a cucumber, but melons like other melons, squash like other squash, and cucumbers like other cucumbers. Because these plants flower throughout the season, you'd have to limit yourself to planting just one variety at a time to ensure genetic integrity. Tomatoes, beans, peppers, lettuce, and peas don't cross as readily and are therefore far simpler choices for the beginner seed saver.

Hand-pollinate if you're up for the challenge. Female flowers can be placed in pollination bags to protect them from receiving unwanted pollen. Remove the stamens (the male flower parts covered in pollen) and transfer the pollen to the female flower. Repeat this for three to five days. Leave the flower protected within the pollination bag to ensure it doesn't receive additional pollen from an unwanted male flower. Remove the bag once the flower drops and the fruit begins to develop.

Let the pods or fruits fully mature. Lettuce, brassicas, carrots, dill, most flowers, onions, and spinach are examples of plants that form seed heads. Allow the seed heads to become dry and brittle before harvesting. Bean pods are treated similarly. Don't leave them out too long or they may burst open or the seeds may start to sprout within the pods. Allow fruits such as cucumbers and zucchini to become extremely

I plant just a single variety of arugula for seed saving, which eliminates the need to space varieties far apart.

large and develop a tough outer skin. Melons, tomatoes, winter squash, and peppers are harvested for seed saving when in their edible state.

Let seeds dry indoors a bit longer. Spread out the seeds and pods on a plate and let them dry for about a week longer after harvesting. Seeds should be as dry as possible to prevent disease or premature sprouting.

A NOTE ABOUT BEANS

Beans are the simplest seed to save because they rarely cross-pollinate and it's easy to tell when pods are mature. Save seeds from any variety of bean including pole beans, bush beans, lima beans, fava beans, and even soybeans. Save the biggest pods with the greatest number of beans per pod for use as seed. Selecting the best of the crop for years can yield a bean that has been refined for your growing conditions. Be sure that the seeds are dry and transfer them to the freezer for 48 hours to kill any bean weevil eggs.

A NOTE ABOUT BRASSICAS

Brassicas are represented by just a few species of plants despite their diverse growth habits. Members of the same species can readily cross-pollinate if within 800 feet to ½ mile of one another. The plants are biennial, meaning that they produce a flower in their second year of growth. However, my brassicas reliably produce flowers during their first year once temperatures begin to rise in spring and early summer.

When saving brassica seed, ensure that you're allowing only one variety from each species to flower at a time (see the variety lists at right). For example, you can't save seed from collards that were flowering alongside broccoli or cabbage. However, 'Georgia Southern' collards (*Brassica oleracea*), mizuna (*B. rapa*), and Ethiopian blue kale (*B. carinata*) can all flower simultaneously and the seeds will possess the same traits as the parents. Staggering the plantings or hand-pollinating can increase the number of reliable varieties you can save each growing season.

Here are a few easy-to-grow varieties from each species.

Use pollination bags for insect-pollinated flowers such as zucchini to ensure true-to-type seed when growing multiple varieties capable of cross-pollinating.

B. oleracea

- Broccoli
- Brussels sprouts
- Cabbage
- Cauliflowers
- Chinese cabbage
- Collards
- Kohlrabi
- Lacinato kale
- Romanesco
- Scarlet kale

B. rapa

- Broccoli raab
- Mizuna
- Pak choi
- Tatsoi
- Turnips

B. napus

- Russian kales
- Rutabagas
- Siberian kales

B. carinata

- Ethiopian blue kale

B. juncea

- Mustards

A NOTE ABOUT SQUASH

Four species of squash, pumpkins, and gourds—*Curcurbita maxima, C. argyrosperma, C. moschata,* and *C. pepo*—provide us with a plethora of diverse varieties. With very few exceptions, summer squash and zucchinis are *C. pepo*. Winter squash and pumpkins can be one of any four species. As with brassicas, different types of the same species (such as 'Blue Hubbard' and candy roaster, both *C. maxima*) can easily cross-pollinate, and the offspring is often an inedible gourd that's nothing like the parent.

Choose just one squash variety from each species to flower at a time. For example, seeds could be saved from 'Red Kuri' (*C. maxima*), cushaw (*C. argyrosperma*), 'Black Futsu' (*C. moschata*), and a single variety of zucchini (*C. pepo*) each year. Hand-pollinate or stagger flowering times to save seed from more varieties. Luffa gourds are their own species and can't cross with any other cucurbits; their seeds can therefore be saved unless you plant multiple luffa varieties.

Here are some common varieties from each species, designated as summer (s) or winter (w) squash.

C. maxima

- 'Blue Hubbard' squash (w)
- Buttercup squash (w)
- Candy roaster squash (w)
- Crown pumpkin (w)
- Kabocha squash (w)
- 'Red Kuri' squash (w)
- 'Turk's Turban' squash (w)

C. argyrosperma

- 'Black Sweet Potato' squash (w)
- Cushaw squash (w)

C. moschata

- 'Black Futsu' squash (w)
- Butternut squash (w)
- 'Long Island Cheese' pumpkin (w)
- 'Tromboncino' squash (w/s)

C. pepo

- Acorn squash (w)
- Crookneck squash (s)
- Delicata squash (w)
- Pattypan and scallop squash (s)
- Spaghetti squash (w)
- Straight neck squash (s)
- Zucchini (s)

'NODAK' PINTOS   'LAZY WIFE' BEANS   'CHEROKEE TRAIL OF TEARS' BEANS   'CHOCOLATE' RUNNER BEANS

top: I save seeds from most of my bean plants every year.

below: A flat of overwintered mizuna quickly bolted in the spring and will be used for seeds.

①

Fermenting Seeds

Seeds from cucurbits and nightshades (tomatoes, peppers, ground-cherries, eggplant, and so on) can benefit from a one- to two-day ferment immediately after harvest.

②

1. Collect the seeds and pulp from a selection of ripe fruits, ideally borne on different plants, in a glass jar. Here I'm saving seeds from Golden physallis, a sweet fruit resembling a tomatillo. The plants were separated from tomatillos by 800 feet to prevent cross-pollination.

2. Cover the seeds and pulp with a couple of inches of water and allow them to ferment for a day or two. The viable seeds will sink while the pulp remains suspended.

3. Remove the pulp and floating seeds, if any, then rinse the remaining seeds and strain them through a coffee filter. Tear the filter open and lay it flat so the seeds can dry. Place the fully dried seeds in paper envelopes and store them in a dark spot at room temperature.

③

PLANTING DISTANCE FOR SEED SAVING

Use the following chart to inform your seed-saving efforts. Plants that require long distances between varieties can be saved if you isolate pollen to selected flowers. For example, planting a single variety or allowing just one variety to flower at a time eliminates the need for distance requirements.

Placing flowers in pollination bags (see page 182), or even paper bags or cheesecloth bags for large flowers such as corn, protects them from receiving unwanted pollen. Protected flowers must be pollinated by hand.

Finally, you can always save seed from cross-pollinated varieties; just don't expect an offspring that's identical to the parent. In some instances, as with zinnias and sunflowers, the result is still a pretty flower. In others, such as corn, you're better off buying new seed rather than planting a cross-pollinated variety.

PLANT	DISTANCE BETWEEN VARIETIES
Arugula	800 feet–½ mile
Beans, bush and pole	10–20 feet
Beans (fava and lima)	50 feet
Beets	½ mile
Brassicas	800 feet–½ mile
Carrots	800 feet–½ mile
Celery	½ mile
Chard	800 feet–1 mile
Corn	At least ¼ mile; 1 mile is best
Cucumbers	800 feet–½ mile
Eggplant	300–1,600 feet
Endive	50 feet
Lettuce	10–20 feet
Melons	800 feet–½ mile
Okra	500–1,600 feet
Onions	800 feet–1 mile
Peas	10–20 feet
Peppers	300–1,600 feet
Radishes	800 feet–½ mile
Spinach	½ mile
Squash (summer, winter, pumpkins, and gourds)	800 feet–1 mile
Tomatillos	800–1,600 feet
Tomatoes	10–50 feet

farm-fresh
EGGS

Eggs from your own hens are packed with valuable proteins and fats, two of the macronutrients less easily found in food crops and important to human health. Chicken manure introduces a rich source of nutrition to garden soil. These diverse and important benefits of owning chickens are accompanied by a major commitment: Raising animals is demanding in that procrastination can equal neglect. Failing to sow seeds during an optimal window is excusable; allowing chickens to run out of water is not. If you're up for the commitment, raising your own chickens is deeply rewarding as well as a source of ongoing entertainment!

A movable chicken tractor serves as a temporary home for the chickens atop a new garden bed. They'll spend several days removing weeds and bugs, and adding nitrogen-rich fertilizer.

the plan

- **BUILD OR BUY A COOP.** Chickens need to roost at night in a predator-proof enclosure equipped with egg boxes for laying. They need plenty of ventilation but require supplemental heat only in extremely cold climates. The hardiest breeds can withstand subfreezing temperatures if the birds remain protected from wind and moisture.

- **DECIDE WHERE THE BIRDS WILL SPEND THEIR DAYS.** Chickens can be enclosed within a stationary pen called a run, moved to fresh grass daily in a movable coop known as a chicken tractor, confined within a movable electric fence that's rotated weekly, or allowed to free range. Each of these options comes with its own set of benefits and challenges, so weigh all your options before committing to a technique.

- **BUY PULLETS FOR AN EASIER START.** Brooding chicks is messy business, requires additional equipment, and demands a great deal of attention to keep the babies alive and well. Many local farms offer pullets (female) and cockerels (male) that are 8 to 16 weeks old, fully feathered, and ready to go in an outdoor coop. The higher price paid for older birds is partially offset by the savings in equipment, feed, and time you would need to raise chicks.

BEFORE YOU DIG IN

There's nothing quite like a fresh egg. The anticipation and excitement of gathering fresh eggs from the coop every day are feelings that never get old. Raising your own chickens will provide you and your family with some of the tastiest eggs you've ever had, and you'll have peace of mind in knowing exactly how that bird was raised. However, there are a few things you need to consider.

You're unlikely to save money. Truth be told, there's little to no cost savings to be had by raising your own layers, especially if you're buying organic feed. Upscaling the size of your operation, buying feed in bulk, and selling eggs at a premium can help cover some of the expenses, but the investment in materials, birds, and feed doesn't quickly pay for itself.

Check town, city, or HOA restrictions. Ordinances often limit the number of birds you're allowed to keep, how they can be kept, and whether you can have a rooster. Look into permits before starting construction.

Chickens destroy gardens. As efficient diggers and foragers, chickens will uproot plants, destroy mulching efforts, and eat everything in sight. If you wish to free range chickens, you need lots of space between you and your closest neighbor plus fences around your gardens.

Chickens like to roam. Chickens will wander away from your house into areas where they can encounter predators or bother neighbors. Some breeds are more likely to seek out adventure; others are more the homebody type. Chickens can be coaxed back into their pen using treats such as dried mealworms or cracked corn, but they often catch on to the trick and choose to remain free.

Our kids are far more likely to show our ducks and chickens to their friends than they are to point out any other feature on our homestead. We may not be saving money, but the value in raising birds extends far beyond eggs.

The pecking order is real. Chickens establish a natural order of dominance over one another. If a chicken leaves the flock or a new bird is introduced, that order must be reestablished. Even if you meet the recommended space requirements to prevent squabbling, know that some pecking and squawking is to be expected from time to time. Intervention may be necessary if a chicken begins to suffer wounds or is being prevented by other chickens from eating or drinking.

Predator prevention is key. Weasels, coyotes, raccoons, hawks, opossums, dogs, foxes, and cats are interested in eating your birds, while skunks, squirrels, crows, and other scavengers are interested in eating their feed. You can't stop them from coming around, so design your facilities knowing that predators absolutely must be controlled. Don't wait until a problem arises to predator-proof the structure. Once a predator finds a reliable food source, it is likely to return.

MORE THAN JUST EGGS

Regenerative farming practices aim to improve the quality of soil over time using five principles:

- Keep the soil covered year-round with living roots or mulches.
- Diversify to enhance ecosystem stability and productivity.
- Allow roots to grow in the soil year-round.
- Minimize soil disturbances (no-till gardening).
- Integrate livestock.

You're likely already implementing many of the soil-management techniques recommended by regenerative farmers. Incorporating chickens into the garden allows you to focus on the final principle—integrating livestock. Cows, sheep, and other grazers stimulate the movement of carbon into the soil through grazing, but keeping such large animals is impractical for growers with limited space. Chickens offer some of these same benefits but without the demand for large acreage.

Chickens produce manure. Farm animals efficiently transform plant matter into an abundant source of nutrient-rich manure that can be a boon in a gardening or farming system. Poultry manure contains all 13 essential plant nutrients—nitrogen, phosphorus, potassium, calcium, magnesium, sulfur, manganese, copper, zinc, chlorine, boron, iron, and molybdenum—and can serve as the sole supply of plant nutrition in a garden.

To capture this nutrition, you must have a few inputs, including animal feed and an abundant source of carbon to absorb the animal waste (see Set Up a Deep-Litter System, page 206), but with the trade-off come fresh eggs and loads of nutrients for your garden.

Chickens eat bugs and weeds. A chicken tractor (page 207) can be used to isolate birds to a particular bed where they'll scratch, consume bugs and larva, deposit waste, and remove weeds. This is a great technique to use when a crop is spent and you want to boost soil nutrition and to remove pests, weeds, and debris before planting the next crop. Releasing a flock into the garden without the confinement of a chicken tractor can be catastrophic, however. The diligent work of one or two chickens may go unnoticed, but the work of even three or four can be significant,

DEATH IS PART OF LIFE ON THE FARM

>>> While infrequent, there are instances in which a chicken will become sick or injured and start to noticeably decline. A vet visit can be expensive—if you can even find a vet who will examine a chicken. At times, the most humane choice is to end the suffering of the bird rather than let the disease run its course, spread disease to the rest of the flock, or allow prolonged suffering from injury. Furthermore, poultry egg production drops by about 10 to 20 percent every year. Will you continue to feed and care for your birds when the cost of feed begins to outweigh the benefit of egg production?

Always implement strategies to keep birds safe and healthy but know that the loss of one or two birds every year, and sometimes more when dealing with chicks, is to be expected. How you plan to handle sickness, injury, and death should be a major consideration in your commitment to raising animals. Talk with other growers and farmers, and search online to learn more about these special considerations while planning your egg-production efforts.

especially in a small garden. They'll remove everything, not just the weeds.

SET REALISTIC EGG-SPECTATIONS

Some chickens start laying as early as five months old, but it's far more likely that their first eggs appear somewhere between the ages of six and eight months Larger breeds generally take longer to start laying. Eggs from younger birds are small but will slowly increase in size. As hens age, they lay fewer but larger eggs. Expect a 10 to 20 percent decrease in the number of eggs produced per chicken every year. Many chicken breeds will stop laying altogether at six or seven years old. The bizarrely named Deathlayer chicken is the only breed to my knowledge that will lay for a lifetime.

Daylight stimulates reproductive hormones, so chickens lay very few eggs in the winter when daylight drops below 10 to 12 hours per day. Supplementing with artificial light will improve production, but I don't recommend it. Egg production wanes during a time of year that chickens need the energy to replenish their protein reserves after molting (see page 213) and to keep warm. The extra strain of egg production is unnatural and places too high an energy demand on birds.

You need to collect the eggs daily. A pile of eggs left uncollected will encourage broodiness (see page 215). And while not common, chickens will eat their own eggs—especially if piles of eggs sit in the nesting boxes for days—which can be a sign of nutrient deficiency. To correct this, collect the eggs more frequently (up to three times per day) and supplement the flock's diet with dried mealworms or soldier fly larva.

Eggs come in a much wider variety of colors than you can find at the grocery store. In addition to white and light brown, they can be pale pink, chocolate brown, green, or even blue. A given breed of chicken typically lays a predictable color of egg.

PLAN YOUR COOP CAREFULLY

When designing a coop, think about what it'll be like to clean it. Will it require that you twist your body into strange and unnatural positions to reach back corners? My coop is difficult to clean, and I'm anxiously awaiting the day that it becomes decrepit enough to justify replacing it. Most first-time chicken owners do exactly what I did, which was to build my coop using scrap materials pieced together in whatever fashion I could.

I agree with this approach in that chickens are a costly investment and it's wise to guarantee a commitment and build knowledge before spending large sums of money on coop construction. However, always keep maintenance (including cleaning) at the forefront of your plans.

Most wild predators are nocturnal, meaning that a well-built coop can successfully fend off most threats. Aerial daytime predators such as hawks require alternative protection measures including a shelter directly over the birds' feeder or a bird-netting cover over the entire run. Chicks and juvenile birds are the easiest targets and should be kept in covered enclosures until they're fully grown.

The following housing options—stationary coops, chicken tractors, movable electric fencing, and free ranging—all come with their own benefits and challenges, and each requires a different set of inputs. When managed correctly, the following methods will provide your animals with exactly what they need to stay safe, healthy, and productive.

I carefully pieced together scrap wood to meet the space, roost, and nesting box requirements for a flock of 12 chickens. The coop serves its purpose but it's extremely difficult to clean—a factor I wish I had given more attention to.

Unlike region-specific predators, dogs are ubiquitous. They are active during the day and often allowed to roam in rural areas. Installing proper fencing or owning a guardian dog are the only ways to prevent dog attacks.

CHICKENS ARE MESSY

Chickens have many charms, but they dig and poop A LOT.

>>> Chicken mess results from the birds' insatiable desire to dig and frequent need to poop. They'll dig in garden beds, mulched areas, lawns, borders, and any area filled with underground critters to feast upon. They'll poop on patios, lawns, gardens, the backs of the neighbor's lawn chairs where you didn't know they were roosting all afternoon, and anywhere else they please.

Allowing birds to free range sounds lovely in theory; in practice, it results in more problems than most people are willing to tolerate. For this reason, most chicken owners opt for runs, fences, and other means to at least limit their chicken's access to certain areas (see page 209).

STATIONARY COOP WITH ATTACHED RUN

Most backyard chicken owners, especially ones with limited space, house their birds in a stationary coop (house) with an outdoor run (pen) for ease of construction and maintenance. Any lawn or ground cover within the run will quickly be destroyed by the digging action of chickens and will stink if not properly managed. Stationary runs should be maintained using a deep-litter system (page 206) to keep the run clean and chickens healthy. My properly managed deep litter generates nearly all my compost needs, which is my primary reason for keeping my chickens within a stationary coop and run.

PROVIDE ENOUGH SPACE

Overcrowding is the leading cause of fighting and pecking among birds. Downsize if the flock seems too crowded, but it's best to prevent the squabbling through proper coop and run design. Although tempting, never house more birds than your space can comfortably accommodate.

Floor space. Each chicken needs 2 to 4 square feet of floor space. Although chickens spend more time on their roosts than on the floor, they still need ample space to move around.

Ventilation. Chickens have sensitive lungs that can be damaged by the buildup of gases released from their waste. Provide at least 2 square feet of ventilation per 10 square feet of floor space. Using this calculation, a 30-square-foot coop needs 6 square feet of ventilation. Keep vents open year-round, even in winter. Cover all windows, large cracks, and vents with ½-inch hardware cloth, which offers better security than chicken wire.

The coop serves as a predator-proof enclosure while the run is reinforced with two layers of wire to keep out four-legged visitors and shade cloth to keep out aerial predators.

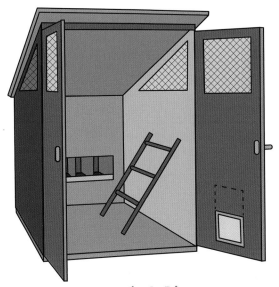

This 5 × 7-foot coop design can easily house 10 to 15 chickens.

Roost space. Chickens roost off the ground at night; they need 8 to 10 inches of roost space per bird. Chickens are flat-footed, so they prefer a rectangular roost with rounded edges. They like to curl their toes around the roost at the front and back, so use a board with a width of about 2 inches. Roosts should be:

- 1 to 3 feet above the ground (any higher and the chickens will need steps or a ladder; lower and they may not feel as if they've found a secure place to roost)
- 18 inches or more away from a wall or ceiling
- 18 inches or more away from another roost

Angle the roosting ladder so that droppings won't fall onto a chicken below. Secure it in place using screws.

Nesting boxes. Provide one nesting box for every three or four chickens. If possible, place the nesting boxes off the ground to keep the hens feeling safe. Create a 1 × 1 × 1-foot box or series of boxes that are large enough to hold a

SOURCING MATERIALS

>>> Coops can be constructed out of an array of repurposed materials ranging from old doors and plywood scraps to an upgraded old shed. My coop was built using plywood from an old climbing wall and stones. A friend built hers from scrap metal and a retired trampoline. Search for coop construction materials like old doors, windows, and other scrap materials at used building-supply stores. Creativity goes a long way in saving money!

There's no right or wrong way to build a coop as long as it keeps the birds safe and healthy. If time and/or building skills are limited, look for kits or preassembled coops at farm supply and hardware stores. Note that kits shouldn't always house as many birds as advertised.

An old trampoline and scrap metal were creatively pieced together to build this backyard coop.

chicken while keeping it protected and secure. Hens won't lay in boxes where they feel that their brood may be vulnerable to attacks.

BUILD THE COOP

Not all coops have floors; some are built directly on bare ground. Fencing should be buried along the bottom perimeter of the coop to prevent predators from digging under it. Coops with a floor are generally more secure against predators as the entire coop is impenetrable once the door is closed. If building a floor, line it with linoleum or stick-on tiles to make for easy cleanup. Keep the floor covered with 4 to 6 inches of wood shavings and replace the shavings every week.

My nesting boxes are tucked into the back of the coop to help the chickens feel secure. The boxes can be removed for easy cleaning.

MEET YOUR NEW COMPOSTER: CHICKENS

>>> I often liken chickens to a garbage disposal in that they'll eat almost anything. Some chicken owners are more diligent about what they feed their birds, but I'm of the mindset that nearly all compostable items, excluding junk food and the items listed below, go into the chicken run.

WHAT NOT TO FEED TO CHICKENS

- Highly processed foods
- Nightshade leaves (potatoes, tomatoes, peppers, and tomatillos)
- Caffeine and alcohol
- Uncooked beans, pasta, and meat (cooked versions are okay)
- Ornamental landscaping debris (vegetable garden debris only)

You can find far more extensive lists of foods that chickens shouldn't eat, such as avocado skins and pits, potato peels, and chocolate. I largely ignore these suggestions, assuming that chickens will instinctually know what they can and can't eat. Uneaten material gets worked into the deep litter for composting.

Chickens happily tuck themselves in for the night when the sun starts to set and will emerge from the coop in the morning when it's light. They'll do this of their own accord without force or encouragement. For this reason, I recommend installing automated coop doors that operate on timers or light sensors to open and close accordingly. While automated doors aren't a requirement, forgetting to close a coop door after the chickens have put themselves to bed just once can be catastrophic if a nighttime predator slips in. Automating the system is a foolproof way to keep the flock safe at night while giving you the flexibility to leave the flock unattended for a few days.

BUILD THE RUN

Each chicken needs a minimum of 15 square feet of space within the run to prevent squabbling and overcrowding. Most sources recommend 10 square feet of space per chicken, but this has never felt adequate to me; more is better. See Free-Range Birds, page 210, if you're trying to maintain grass within the run; otherwise, plan to manage the waste using a deep litter system (page 206).

Make runs as predator resistant as possible. Nocturnal predators like raccoons, opossums, and foxes can climb into the run unless it's completely enclosed with a roof. This type of extensive construction isn't always practical or affordable, so the goal with a run is to keep the birds in and the daytime predators, like dogs, out. The coop should reliably protect against any critters that wander into the run at night.

Four-foot-high fencing made of 2 × 4-inch welded wire is sufficient in most cases, but 5-foot fencing offers a bit more protection and helps prevent the chickens from flying out. Add a layer of 1-inch chicken wire if you find that smaller

animals like skunks can squeeze through the welded wire fencing at night. Attach the chicken wire to the outside of 8-foot-long, 4 × 4 wooden posts with fencing staples or to metal U-posts using zip ties or metal wire.

I suggest using 8-foot wooden posts (bury 2 feet of the post in the ground) so that the posts are taller than the fence, making it possible to stretch bird netting or shade cloth over the top while giving you room to walk underneath. (Use 10-foot posts if you're taller than 6 feet.) This added layer provides shade and prevents the

My neighbor's coop with an open dirt floor is tucked into a corner of the garden and attached to a run via a small wire tunnel. The upper level is protected from the weather and houses the nesting boxes.

chickens from flying out and predatory birds from flying in. Drive nails into the tops of the posts, stretch the shade cloth or bird netting over the run, and hook it onto the nails. Use bird netting in especially snowy climates as the holes are larger than shade cloth and snow is less likely to accumulate on top.

Chickens like to roost in the run throughout the day. Give them plenty of options when it comes to roosts and perches so that they're able

Shade cloth and bird netting draped over the top of 4 × 4 posts provide shade and prevent chickens from flying out and predatory birds from flying in.

CLIPPING YOUR BIRDS' WINGS

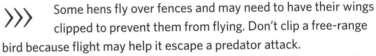 Some hens fly over fences and may need to have their wings clipped to prevent them from flying. Don't clip a free-range bird because flight may help it escape a predator attack.

To clip a bird's wings, remove the outer 3 inches of the longest flight feathers, just below a set of smaller feathers. It's a quick and painless process as long as you don't cut the feathers too short. Clip just one wing to throw the bird off balance; she'll be unable to fly up and over fences afterward. Be careful not to cut the thick shaft of the feathers or the bird can experience heavy bleeding.

Clipping is a two-person job. One person holds the chicken's body, holding one wing still and supporting the feet with their other hand. The second person extends the free wing and removes the feathers using sharp scissors. Always clip less when in doubt. Going back to clip more is always an option if you find that your efforts were ineffective. Repeat this process once per year because birds regrow feathers annually after molting.

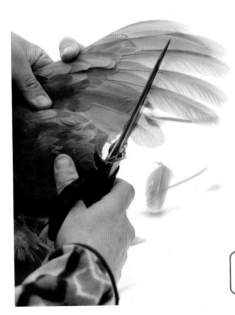

Clipping wings is a simple task but one that requires two people.

to get a bit of flight exercise. Roosts can also serve as a refuge for chickens that are being bothered by a flock mate.

BUILD ADDITIONAL SHELTER

Chickens have limited ability to cool themselves when it's hot. Like dogs, they'll pant. In warm weather, place the coop and run in the shade or cover it with shade cloth to prevent overheating. During the rest of the year, provide a covered area where chickens are protected from harsh sun, rain, snow, and wind. Chickens don't like to be wet, so they need a covered space where the entire flock can congregate when it rains.

Chickens keep clean by taking dust baths to exfoliate dead skin, shed worn feathers, and smother parasites such as lice and mites. They dig holes in dry dirt, filling their feathers with dirt as they dig. They fluff their feathers, sprawl out their wings, and allow the dirt to sit for quite some time before shaking it from their bodies.

You can create a dust bath by placing a kiddie pool (or other wide, shallow container) filled with dirt underneath a tarp or shelter. Sprinkle a very thin dusting of diatomaceous earth or wood ash on top of the soil every month or so to further protect the birds from parasites. The important thing is that the area remain dry so that the chickens have a place to bathe.

A small awning attached to the dust bath area gives the chickens a covered space where they can congregate during inclement weather.

A large coop door covered in hardware cloth offers plenty of ventilation. This battery-powered automatic door senses light and opens and closes accordingly. Other models can be programmed to open and close using a timer, but they'll need to be adjusted seasonally to reflect changes in daylight length.

1 Add 12 to 18 inches of carbon (brown matter) to the run: wood shavings or baled pine bedding work well, as do dried leaves (mixed with other materials to avoid becoming waterlogged). Wood chips take much longer to decompose, so if you use them, remove the litter from the run after two to three months and allow additional time for complete decomposition (1 to 2 years depending on particle size).

2 Let the chickens work their composting magic as they scratch and dig in the litter, turning in their nutrient-dense excrement all the while. Worms, grubs, and other decomposers will start to grow in the litter, providing a valuable source of food for the chickens and further encouraging them to turn the compost.

3 Throw kitchen scraps into the run. The chickens eat whatever they choose, then mix in the rest with the contents of the deep litter to quickly decompose. See Meet Your New Composter: Chickens on page 202 for more information on what can be fed to chickens.

4 The frequency of litter collection depends on the size of the run, number of chickens, and climate (wet climates require more frequent cleaning due to more rapid decomposition). Collect the deep litter every few months using a 10-tine compost fork and wheel barrow when most of the carbon particles have decomposed into rich compost. You've waited too long if the run starts to stink, a sign of nitrogen off-gassing. Remove the litter immediately if you notice any foul smells. Leave approximately 20 percent of the litter in the run to jump-start the decomposition process again.

5 Let the litter sit for a few months before applying it to the garden. If need be, you can transport the litter directly to the garden beds, but it's best to let the litter sit for two to six months to allow for more complete decomposition.

STEP-BY-STEP
Set Up a Deep-Litter System

Poultry manure is extremely high in nitrogen. The soil within the run can absorb that nitrogen up to a point, but the excess transforms into ammonia, nitrous oxide, and other forms of nitrogen gas that have negative impacts on the environment (this off-gassing is responsible for smelly manure). Managing and capturing all that nitrogen requires the addition of carbon— deep litter—which benefits both the garden and the environment.

With a deep-litter system, a bigger run isn't always better, as you're moving large amounts of organic matter on a regular basis. Limit yourself to no more than 15 chickens in a run no larger than 300 square feet to avoid becoming overwhelmed. Ten chickens within a 200-square-foot run is my preferred ratio. More than that and I find it difficult to keep up with the amount of waste generated by my flock. Growers interested in raising more than 15 birds should explore movable electric fencing instead of a run.

MOVABLE CHICKEN TRACTOR

Chicken tractors are predator-proof enclosures that allow you to move the birds to fresh grass daily while utilizing a small footprint. With this system, the tractor must be mobile—build it on wheels or make it light enough that it can be dragged. Tractors are often used for dense populations of meat birds who spend just a few months in the tractors before being processed, on the theory perhaps that living in close quarters isn't such a big deal when life is so short.

When it comes to layers, tractors are useful for those interested in keeping just a couple of birds. To prevent excessive and sometimes damaging squabbling and to ensure bird health, never house more than two layers in a 40-square-foot tractor, four in an 80-square-foot tractor, or five in a 100-square-foot tractor. Growers wanting more than just a few birds should explore other methods.

Keep the design light and mobile. Chicken tractors should be moved to fresh grass daily or weekly depending on the footprint of the tractor and the number of birds within. Select light materials and place wheels under the heaviest part of the tractor.

The space requirements for stationary coops (page 200) apply to chicken tractors, especially if you plan to use the tractor as a permanent enclosure. Tractors used as temporary enclosures (that is, atop beds for garden cleanup or as an intermediate enclosure for chicks or pullets) don't have to follow the recommendations as strictly as the chickens won't be spending their entire lives in the tractor. Many tractors are built in an A-frame shape for simplicity, but you can build any shape that meets the requirements for roost positioning and size, run square footage, nesting boxes, and ventilation.

My chicken tractor includes two roosts, two nesting boxes, and an area covered with sheet metal to provide shelter. This small structure serves as an intermediate enclosure for raising young birds and an enclosure that can be situated atop spent garden beds to aid in cleanup.

Slip off the wheels using a wing nut or bolt so that the frame fits tightly against the ground when not in use to prevent the entry of predators.

A HYBRID IDEA FOR ENCLOSED CHICKENS

>>> Chickens need a minimum of 300 square feet per bird to prevent them from destroying the grass within the run. This requirement is easy to meet when you need to contain just two or three birds, but constructing a large enclosure for 10 or more chickens can be challenging and expensive. Consider building two runs—a smaller one managed with a deep-litter system and a second one on grass. Attach the second run to the smaller run and make it as large as possible. Limit birds' access to the larger run so that they're able to spend time ranging freely without destroying the grass.

For example, I attached a 1,500-square-foot run to a 200-square-foot run. Ten or so chickens are released into this space for 30 to 60 minutes every evening. They'll return to their coop on their own, so I time their release around sunset. Because they tend to dig in the same spots, I periodically block their access for a week or two to let the grass rebound. I also block access more frequently in winter when the grass is dormant and can't recover as readily. This hybrid system isn't as good as true free ranging, but it offers more exercise and foraging opportunities than the chickens would get otherwise.

MOVABLE ELECTRIC FENCING

Containing your chickens within a movable electric fence with a mobile coop is one of the best methods for protecting the birds from ground predators while allowing access to fresh grass. The many benefits of electric poultry netting include the following.

- consistently provides the birds access to fresh foraging ground
- avoids the need for extensive permanent fence construction
- suppresses parasite transmission through contaminated fecal matter
- improves the quality of the grass because it's fertilized by the birds

Electric poultry netting is practical only when you have enough space for rotation. Fences are generally moved to fresh grass weekly, but the frequency depends on the quality of the lawn or pasture and the density of your flock. Never leave the birds in any one space long enough to destroy the grass. And never move them back to a repeat location until the grass has fully rebounded.

Netting. Electric poultry netting is available in a standard length of 164 feet or half-length of 82 feet. The ends of the fencing don't have to meet, which means that an existing structure such as a building can be integrated into the fence line. Most people use 42-inch-tall netting, but 48-inch netting may be necessary to protect birds from coyotes or dogs. Longer, taller netting is more expensive and harder to move.

Power supply. Fence energizers to provide electricity to the fence can be powered by batteries, solar units, or A/C outlets. Solar-powered units are the most costly but offer the most flexibility. The initial cost is often recouped over time by eliminating the need to replace batteries. The size of the energizer required to power the fence depends on the length of fencing and the height of the grass surrounding the bottom of the fence. Mowing the grass along the bottom of the fence line allows energy to move more freely along the length of the netting. Contact an electric fence supply company such as Premier 1 Supplies before purchasing a kit to be sure you choose one suitable for your situation.

The electric fence offers an excellent protective barrier for nighttime predators, but it's best to take steps to predator-proof the coop as well. Follow the construction recommendations listed in Stationary Coop with Attached Run (page 200).

Electric fencing is a convenient way to contain your flock while rotating them to fresh grass.

FREE-RANGE BIRDS

Allowing chickens to free range is every egg seeker's dream. Birds are certainly happiest and healthiest when raised this way, as they get ample exercise and unlimited foraging access. But free-range birds are the most susceptible to predation, and they can be a nuisance. Free ranging is therefore only possible in certain situations, but even then, it's risky business. If you're set on free ranging, consider the following suggestions.

Fence around your property borders to keep birds in and daytime predators out. Predator-proof coops (built to the specifications listed for stationary coops on page 200) are still necessary, because many nighttime predators can climb up and over fencing.

Fence around gardens to keep out birds. Use additional fencing to limit the areas where the birds can free range.

Give them room. Even large fenced-in areas are susceptible to chicken damage. Provide each chicken with at least 300 square feet of space, more for poor-quality grass. For example, three chickens require a minimum of 900 square feet if you're to maintain lawn within the fence.

Supervise free-ranging birds. Don't let your birds wander around when no one is home to keep an eye on them, and respond to any signs of distress.

PROVIDING FOOD AND WATER

Chickens don't need a lot of attention, but they do need regular care. Use automated systems for feeding, watering, and even the daily opening and closing of the coop door to simplify your chores, not to foster negligence. Meeting the chickens' daily needs will keep the birds alive and well and ensure you have plenty of fresh eggs.

Chickens are prone to dehydration, especially when it's hot. Waterers come in varying sizes ranging from 1 to 7 gallons or more.

Large basins for supplying water are available in multiple designs. The waterer shown here is one of the most common designs—search online to see what would best meet your needs. Some waterers attach to a hose for a continuous supply of water; others are equipped with heaters to prevent freezing in winter.

Five gallons will keep 10 chickens hydrated for about a week. Some sources suggest adding a small amount of bleach to the water to keep it clean, but this can damage the birds' gut flora. Instead, clean the waterer frequently, and add 1 tablespoon of apple cider vinegar per gallon of water to prevent algae growth. Elevate the waterer 10 to 12 inches off the ground so that the chickens don't kick dirt and debris into the trough.

A single chicken eats about 1 pound of feed per week. Layer feed is specially formulated to provide chickens with a balanced nutrient profile of proteins, fats, and carbohydrates. The simplest way to obtain feed is to purchase it from a reputable feed producer, but if you're motivated to make your own feed, you'll find recipes for homemade formulas online.

Crumbles are the most nutrient dense, as pelletized formulas may lose valuable nutrients and enzymes when heated to high temperatures during compression. However, chickens tend to pick through crumbles, taking their favorite ingredients and leaving the rest behind. To prevent this, feed the birds only as much as they'll eat in a day. Or use a rectangular automatic feeder with a screen-covered trough at the bottom to minimize the pickiness. Feed is released only once the area below it is cleared of food.

This Little Giant galvanized high-capacity poultry feeder prevents the chickens from stepping into their food and scratching through it to pick out their favorite tidbits. This design provides continuous feed, unlike some of the cylindrical versions where feed can become stuck.

The chickens must eat all that's available before more feed is dispensed.

Chicken scratch is much cheaper than chicken feed but, like corn, is more of a treat than a meal. Corn and scratch should be limited to no more than 10 percent of the chicken's diet to prevent excessive weight gain.

Free-range birds require less feed at certain times of year. Chickens that are provided with access to fresh grass can forage for much of their own food and therefore require less feed. The exact amount of supplemental feed depends on the season, as more food sources are available in summer than in winter months. Never reduce the feed so much that you see a drop in egg production, a sign of inadequate nutrition.

Deliver greens and bugs to hens in stationary enclosures. Birds confined to stationary runs without access to fresh grass have limited access to valuable food sources like greens and bugs. Bring weeds and other plant debris from the garden to the birds to re-create this natural feeding pattern. If space and climate

CRUMBLES   PELLETIZED FEED

allow, grow chickweed, hairy vetch, grasses, grains, and other winter crops to supply your birds with fresh greens during winter. Dried soldier fly larva, which are generally less expensive and have higher protein content than dried mealworms, are a convenient source of bugs. Provide each confined bird with roughly a tablespoon of mealworms per day. Simply sprinkle the worms on the floor of the run.

Provide birds with grit. Birds don't have teeth, so grit is an essential part of their digestion. They swallow their food whole, where it travels to a collection organ known as a crop. Crops are empty in the morning but full by the end of the day as the bird fills it with food. A lump protruding from one side of the bird's neck late in the day signifies a full crop. From the crop, food moves into the stomach, then into the gizzard, where bits of tiny rock (grit) break the food into smaller pieces. Some brands of commercial feed contains grit, but if you're not using one of those, provide the birds with a small bowl of purchased grit that they can access as they please.

WINTER CARE

Chickens are hardier than you might expect. Some of the larger breeds such as Australorp, Wyandotte, Barred Rock, Delaware, Rhode Island Red, and Orpington can easily withstand subfreezing temperatures without being provided supplemental heat. Small combs and wattles are clues to a breed's cold tolerance as they remain closer to the bird's body and are better protected from cold and thus less likely to experience frostbite.

My unheated coop rarely drops below 5° or 10°F (−15° or −12°C), but I've read accounts of people who don't use heat even when it's −40°F (−40°C). Allowing chickens to acclimate naturally to seasonal temperatures prevents frostbite, because the combs, wattles, feet, and other exposed tissues grow accustomed to the cold. Frostbite is far more common in birds that sleep in warm coops at night and spend their days in freezing temperatures outdoors.

Consider using supplemental heat only when outdoor temperatures consistently remain below 10°F (−12°C). Avoid the use of heat lamps because these have been known to catch fire when operated frequently and continuously. Instead, look for heating pads and other low-wattage heaters specifically designed for use

Hold a chicken's body with one arm while supporting their feet with the other to help them feel more secure when being handled. Note that not all chickens appreciate being handled.

in coops. Alternatively, attaching the coop to a barn or other insulated structure may provide the birds with all the heat they need.

Even in winter, a coop needs to stay well ventilated (see page 200). Don't seal the coop to trap heat, as you will also trap moisture, which is the main reason chickens experience frostbite. Simply guarantee that the birds have a dry place to roost or nest out of the wind, plenty of food so they produce adequate body heat, and ample water. I put my large automatic waterer to rest during winter because it freezes; instead, I deliver a gallon of fresh, warm water in a bowl every morning (sometimes twice a day when temps are particularly low). If you live in a colder climate and are unable to prevent the birds' water from freezing, you may need a water heater.

COMB

WATTLE

CHICKENS MOLT IN LATE FALL

>>> Each year in late fall or early winter, chickens molt: a process in which old feathers are replaced with new ones in preparation for the colder months to come. The birds may undergo a hard molt (during which they lose nearly all feathers and expose large patches of skin) or a soft molt (which can take longer to complete and may not even be noticeable if the chicken loses just a few feathers at a time).

Chickens appear utterly disheveled during the molting process, especially when large patches of bare skin are exposed during cold snaps. Despite their appearance, this is a natural process and my chickens have always survived chilly nights without intervention.

Chickens stop laying when they molt. Feathers are assembled from proteins; during the molt, protein reserves are redirected from egg production toward feather growth. Fortunately this period often coincides with shorter days when hens don't lay much anyway.

Molting chickens typically look kind of awful, but molting is a natural process that occurs every year.

CHOOSE YOUR BIRDS

You can purchase baby chicks (and even eggs for the incubator) from online hatcheries or, better yet, from local farms. While it's certainly entertaining to rear baby chicks, it takes a lot of work, and you'll be waiting up to eight months to get that first egg. Purchasing pullets aged 8 to 16 weeks is far easier, as the birds are fully feathered and ready for outdoor life. There's no need for heat lamps or indoor shelters.

Instructions for brooding chicks can be found on page 219, but I suggest you first check with local farms, Facebook groups, and horticulture clubs such as 4-H or FFA (Future Farmers of America) to see if anyone in your area sells pullets in spring and summer. If buying chicks, consider purchasing them late in spring or midsummer when temperatures are warmer and the birds can safely be placed outdoors sooner.

Introduce new pullets by building an enclosure where the old and new chickens can interact without physical contact for a few days. Everyone can sleep peacefully in the coop overnight, but separate the newcomers from the rest of the flock during the day. Expect some squabbling when you release the new pullets into the existing flock and the birds establish a new pecking order, but the transition will be more peaceful.

Pullets purchased when temperatures are still low will require longer indoor care.

The following considerations are designed to get you started. I've recommended some of the most common breeds, but there are numerous other breeds not listed here. Read the breed descriptions carefully: You'll find that certain traits are more important to you than others. When starting out, select a variety of breeds so that you get familiar with their unique personalities and interesting characteristics. Hone your selection over time once you've found what works for you.

DIFFERENT BREEDS FOR DIFFERENT NEEDS

Layer breeds such as Leghorn, Hamburg, or Ancona are relatively small, lean birds that are laying by five months of age and produce up to 300 eggs per year. Large, fast-maturing chickens such as Cornish crosses are ready for butchering by just 8 to 10 weeks of age and are commonly used for meat production, not eggs. Meat-specific birds could never produce as many eggs.

Barred Rock, Delaware, Orpington, Wyandotte, Plymouth Rock, Sussex, Dominique, Rhode Island Red, and Australorp hens are

DEALING WITH A BROODY HEN

>>> Hens can go broody even when there's no rooster to fertilize the eggs. A broody hen will spend most of her day sitting in a nest whether there are eggs beneath her or not. Trying to remove her from the nest provokes squawking, feather fluffing, and pecking. Removing the hen or isolating her from the nest altogether is in her best interest despite her protests.

Hens can healthfully maintain broodiness for three to four weeks while they await the hatching of eggs, but they'll continue to behave in this way if chicks never arrive at the end of the incubation period. Prolonged broodiness can be detrimental to a hen's health and should be prevented through daily egg collection—a hen that sees eggs piling up in the nesting box is more prone to broodiness. If a hen goes broody and incubating eggs is not in her future, remove her from the nesting box and block her access if possible. Her behavior will normalize in three or four days if properly managed.

considered dual purpose in that they're good layers yet also grow large enough to eat. Dual-purpose birds don't mature as quickly as meat hybrids, but they're a great option for the homesteader who's either interested in raising heritage breeds or aiming to be as self-sufficient as possible and wants both eggs and meat from the same flock. Dual-purpose birds tend to be more docile than the small-bodied layers.

TEMPERAMENT

Different breeds of chickens have somewhat predictable temperaments. There are always outliers, for better or worse, that deviate from expectations, and how you treat your birds is the biggest factor in determining their behavior. Still, always consider a bird's expected temperament when trying to decide if it would be a good addition to the flock.

Buff Orpingtons are one of the friendliest and most docile breeds. As average layers, they're not the best choice for someone trying to maximize production. They are, however, a great choice for someone interested in getting a decent number of eggs from an inquisitive and friendly bird.

Smaller-bodied layers such as Leghorns and Hamburgs are some of the most prolific egg producers, but they're known for being independent, flighty, and relatively uninterested in humans. These birds aren't as well suited for confinement

Some people refer to Buff Orpingtons as the Golden Retrievers of the chicken world because of their people-friendly natures.

Barred Rocks are one of my favorite dual-purpose chickens for their friendly temperament, cold hardiness, and reliable egg production.

in a stationary run compared to some of the larger, less active, dual-purpose breeds such as Australorps, Cochins, or Buff Orpingtons.

Breeds such as Rhode Island Reds have a reputation for being aggressive toward other birds. Mixing docile breeds such as Buff Orpingtons with Rhode Island Reds may result in excessive bickering, especially in a smaller space. Rhode Island Reds and Leghorns are better suited for larger enclosures where the birds have plenty of space to forage and move about. Choose birds with temperaments appropriate for your space.

BROODINESS

Many modern birds have lost the instinct to hatch a clutch of eggs, but breeds such as Australorp, Orpington, Cochin, and Silkie have retained this broody trait. Broody hens become extremely temperamental and stop laying due to hormonal shifts that enable their bodies to drastically reduce food and water intake while they dedicate themselves to sitting atop a nest for three weeks. Broody chickens sometimes even pluck the feathers from their bellies, which allows for direct skin contact with the eggs and thus improved heat transfer. Select breeds known for broodiness if you want your flock to rear chicks.

Ameraucanas are a good dual-purpose breed. They come in a variety of colors and have a distinctive muff of facial feathers.

The Laced Wyandotte is one of the more striking breeds. Plumage comes in gold, silver, blue, or black (pictured here).

CHARACTERISTICS OF COMMON BREEDS

BREED	EGGS/YEAR	EGG COLOR	TEMPERAMENT
Ameraucana	200	Blue	Calm and docile
Australorp	250	Brown	Docile, inquisitive, and calm
Barnevelder	200	Light to dark brown	Calm and docile; even the roosters are known for being friendly
Barred Rock	280	Brown	Inquisitive and docile
Buff Orpington	250	Light brown	One of the most calm and docile
Cochin	170	Light brown	Calm and submissive; broody tendencies
Cuckoo Maran	150	Chocolate brown	Calm and docile
Dominique	250	Brown	Calm and docile
Hamburg	200	White	Active and alert (does not do well in confinement)
Jersey Giant	175	Brown	Calm and docile; even the roosters are known for being friendly
Leghorn	280	White	Intelligent, active, and alert (does not do well in confinement)
Rhode Island Red	260	Light brown	Independent and dominant, somewhat territorial
Speckled Sussex	250	Light brown	Calm and docile
Wyandotte	200	Brown	Calm and docile

WASHING AND STORING FARM-FRESH EGGS

>>> Eggs are covered in a bloom, a protective protein coating that seals the egg and prevents the entry of bacteria. Unwashed eggs can be stored at room temperature for two weeks and in the refrigerator for three months. Egg washing isn't necessary but may be desirable to remove any muck stuck to the shell. Keep the nesting boxes clean to prevent dirty eggs, but grime is sometimes unavoidable, especially during particularly wet periods.

I wash eggs only if they're particularly soiled or if I'm gifting them to a friend or family member. To wash, run the eggs under warm water and use your hands, a dishcloth, or some type of cleaning pad to remove the muck and bloom. The eggs can be left on a clean towel to dry, but once the bloom is removed, they must be refrigerated. Washed eggs will keep in the refrigerator for two months.

To store enough eggs to last through winter when the hens aren't laying, keep unwashed eggs in the refrigerator. Clearly label each batch with a collection date. Rotate the stock throughout fall and winter, using the oldest eggs first and storing newly laid eggs.

Brooders can be constructed from a variety of repurposed materials: an old filing cabinet, a large, sturdy cardboard box, a structure built from old pallets. Whatever you use has to keep the chicks safe, dry, and warm.

HOW TO BROOD CHICKS

Purchasing coop-ready pullets is far simpler than raising chicks, but you may want the experience of raising chicks or have no other option. If this is the case, wait to purchase chicks until the outdoor coop is complete. Birds grow quickly; they'll need that outdoor space before you know it, and I can assure you that you'll be eager to get those new birds settled outdoors as soon as possible.

STRAIGHT RUN VS. SEXED CHICKS

Sexed birds have a known sex at the time of purchase, whereas straight run chicks are those without a known sex. The sex of an unsexed bird won't be known until it matures a bit. Many people prefer to buy sexed chicks so that they can select their desired number of males and females. The issue with buying sexed birds is that males from large hatcheries are often slaughtered by the thousands using inhumane methods.

For this reason, I strongly suggest purchasing straight run chicks. Even better, purchase birds from local farms who, at the very least, have more humane methods of dealing with excess roosters. Even when buying sexed birds, you may end up with a rooster at some point. Will you keep it or give it away? Will it grace your dinner table? Know what you plan to do with that bird when or if the time comes.

SETTING UP A BROODER

Create an enclosure. Select a large plastic tub, galvanized metal basin, or any other enclosure that's at least 1 foot deep and provides 1 square foot of space per bird. Deeper enclosures will prevent escapees as your birds grow older and more agile. Shallow enclosures will need to be covered with a screen or chicken wire. Line the bottom of the enclosure with a couple inches of pine shavings or numerous layers of newspaper.

Provide supplemental heat. Position a heat lamp or heating pad on one side of the brooder to provide supplemental warmth. Always secure a heat lamp to prevent fires, and never allow it to be so close to the birds that they could potentially touch the lamp. Never heat the entire brooder: Direct the warmth to just one side, so that the birds can move to the cooler area if they become too warm. They'll spend less and less time directly underneath the heat as they grow.

A chicken tractor makes an excellent intermediate enclosure for chicks and, in this case, ducklings that have enough feathers to leave the brooder but are not yet large enough to be placed in their permanent enclosure.

IS IT A HEN OR A ROOSTER?

>>> The size of the comb or wattle is often suggested as an indicator of sex but in some breeds, such as Leghorns, hens have large combs that could be mistaken for a rooster's. The most reliable way to determine the sex of a chicken is to observe its saddle feathers—the feathers directly preceding the tail feathers—at three months of age. Roosters boast long, pointed feathers, but a hen's are rounded. A rooster's hackle feathers—the feathers around the neck—are also much longer than a hen's.

If you still find it challenging to distinguish your chicken's sex, wait until they're about four months old and listen for that tell-tale crow.

Notice the rooster's pointed saddle feathers and long hackle (neck) feathers, both distinguishing features of a male bird.

Raise the lamp if you notice that they're continually avoiding the heat.

Use a safe waterer. Baby birds can drown if they are submerged or become dangerously cold if they get wet. Choose a small waterer that prohibits them from stepping into the water.

Prevent birds from standing in their food. Buy a feeder specifically designed to keep chicks from standing in their feed or limit their access to the container. For example, placing a golf ball or baseball in the center of a small bowl of feed often works.

Change the bedding frequently. Chick bedding needs to be changed every four to seven days depending on the number of chicks and the size of the brooder.

TRANSITIONING TO OUTSIDE

The growing chicks will start to spend less and less time under or near their heat source as their feathers mature. Ditch the supplemental heat altogether at four weeks if temperatures remain above 75°F (24°C). Wait until they're fully feathered, around six to eight weeks, if temperatures range in the 50s and 60s (10° to 20°C).

Immature birds are at higher risk of predation and may immediately need a dwelling such as a fully enclosed chicken tractor that can prevent attacks from predatory birds and ground-dwelling threats. Birds can move about freely once they're close to their mature size.

BROODING OUTDOORS

>>> Outdoor brooding can keep the mess of chicks out of your home, but you'll need a draft-free enclosure that is fully predator-proof. I start my chicks indoors but transfer them to an outdoor coop once they start trying to escape the indoor brooder.

Outdoor temperatures need to be above 50°F (10°C) at night for the birds to stay warm enough (even then, you'll need supplemental heat for birds that aren't yet fully feathered). Place the waterer and food bowl above an opening lined with hardware cloth to prevent excess mess. Secure heat lamps to the ceiling of the coop to prevent fires.

Brooders can be set up outdoors if a draft-free garage or other appropriate indoor space isn't available. Don't brood outdoors until temperatures are reliably above 50°F (10°C), otherwise the birds may get too cold.

APPENDIX A: SEEDING RECOMMENDATIONS

Read seed packets for variety-specific recommendations. Some plants such as okra and basil are best started indoors in areas with shorter growing seasons. Plants such as spinach can be succession planted throughout late fall and winter in mild climates. Use this chart as a guide, being aware that you may need to deviate based on climate and variety selection.

CROP	PLANTING DEPTH	SEED SPACING	ROW SPACING	PLANT SPACING	DAYS TO MATURATION	SUCCESSION PLANTING
Arugula	¼ inch	10 seeds/foot	6-8 inches	2-4 inches	30-40	Every 4 weeks when cool
Basil	¼ inch	1 inch	18 inches	8-12 inches	70	N/A
Beans, bush	1 inch	3 inches	18 inches	6 inches	50-80	Every 4-6 weeks
Beans, pole	1 inch	2 inches	24 inches	3-6 inches	80-110	N/A
Beets	½ inch	1-2 inches	12-18 inches	3-5 inches	60-85	Every 4 weeks when cool
Broccoli	¼ inch	Sow in flats	12-18 inches	12-18 inches	75-100	N/A
Brussels sprouts	¼ inch	Sow in flats	24 inches	24 inches	100-150	N/A
Cabbage	¼ inch	Sow in flats	18 inches	18 inches	65-120	N/A
Carrots	¼ inch	15-30 seeds/foot	18 inches	2 inches	70-90	Every 4 weeks
Cauliflower	¼ inch	Sow in flats	18 inches	18 inches	75-110	N/A
Celery	⅛ inch	Sow in flats	18 inches	12 inches	75	N/A
Chard	½ inch	4 seeds/foot	12 inches	8-10 inches	55	Every 4-6 weeks
Chicory & radicchio	¼ inch	8 seeds/foot	12 inches	8 inches	60-80	Every 3 weeks when cool
Cilantro	¼ inch	10 seeds/foot	8-12 inches	2-4 inches	55	N/A
Cress	½ inch	10 seeds/foot	10 inches	4-6 inches	40-50	Every 3 weeks when cool
Collards	¼ inch	3 seeds/foot	18 inches	10-16 inches	85	Every 4-6 weeks
Corn	1 inch	3 seeds/foot	2 feet	12-18 inches	60-110	N/A
Cucumber	¾ inch	4 seeds/foot	4 feet	12 inches	50-70	Every 6 weeks
Dill	½ inch	10 seeds/foot	1 foot	3 inches	90	Every 6 weeks
Eggplant	¼ inch	Sow in flats	24 inches	18 inches	70-90	N/A

CROP	PLANTING DEPTH	SEED SPACING	ROW SPACING	PLANT SPACING	DAYS TO MATURATION	SUCCESSION PLANTING
Kale	¼ inch	3 seeds/foot	12 inches	8–12 inches	55–70	Every 4–6 weeks
Leeks	½ inch	Sow in flats	18 inches	6–8 inches	120–150	N/A
Lettuce	¼ inch	12 seeds/foot	8–12 inches	6 inches	30	Every 3 weeks
Melons	1 inch	2 seeds/mound	4 feet	2 feet	60–110	N/A
Mizuna	¼ inch	1 inch	6–8 inches	4 inches	40	Every 4 weeks when cool
Okra	¾ inch	9 inches	18 inches	18 inches	65	N/A
Pak choi	¼ inch	2 inches	12 inches	8–12 inches	45–55	Every 4 weeks when cool
Parsley	¼ inch	Sow in flats	12 inches	6 inches	75	N/A
Peas	1 inch	1 inch	3 feet	2–4 inches	60–80	N/A
Peppers	¼ inch	Sow in flats	2 feet	18–24 inches	70–110	N/A
Radishes	½ inch	1 inch	6 inches	2 inches	30–50	Every 3 weeks
Rutabagas	1 inch	10 seeds/foot	18 inches	5 inches	70–90	N/A
Spinach	½ inch	10 seeds/foot	8–12 inches	3–5 inches	40–55	Every 3 weeks when cool
Summer squash	¾ inch	2 seeds/mound	5 feet	2–3 feet	50–65	Every 6 weeks
Tomatoes	¼ inch	Sow in flats	4 feet	2–3 feet	70–110	N/A
Turnips	¼ inch	8 seeds/foot	12–18 inches	3–4 inches	50–80	Every 4 weeks when cool
Winter squash	1 inch	2 seeds/mound	10 feet	5 feet	70–120	N/A
Zucchini	¾ inch	2 seeds/mound	5 feet	2–3 feet	60	Every 6 weeks

APPENDIX B: HARDINESS AND SPACING FOR SELECT FLOWERING AND EDIBLE PERENNIALS

The following chart is intended as a guide but is by no means definitive. You may find cultivars that have been bred to be larger or smaller than the most common varieties presented here. The same is true for hardiness zones: Some cultivars can be grown in an atypical range. Use this chart as your starting point, but follow the recommendations on the seed packet or plant tag.

PERENNIAL	PLANT SPACING	USDA HARDINESS ZONES
Asparagus	12–18 inches	2–11
Asters	18–24 inches	3–8
Bee balm	18–24 inches	4–9
Black-eyed Susan	2 feet	3–10
Bloody dock	18 inches	4–8
Chives	18 inches	3–10
Coreopsis	20–24 inches	4–9
Comfrey	2 feet	3–9
Crocuses	6 inches	3–8
Daffodils	12 inches	3–8
Daisies	2 feet	5–8
Daylilies	2 feet	4–9
Delphiniums	18 inches	3–7
Echinacea	2 feet	5–8
Fennel, bronze	2 feet	5–9
Foxglove	2 feet	4–9
Horseradish	3 feet	3–8
Iris	18 inches	3–8
Lavender	18–24 inches	5–9
Liatris	18 inches	3–8

PERENNIAL	PLANT SPACING	USDA HARDINESS ZONES
Lovage	2 feet	3–9
Milkweed	2 feet	4–9
Oregano	2 feet	5–12
Peony	3 feet	3–8
Phlox	24 inches	4–8
Poppies	1 foot	3–8
Rhubarb	3 feet	3–8
Rosemary	2 feet	7–10
Sage	2 feet	5–8
Sedums	8–24 inches (depends on variety)	4–9
Skirret	3 feet	3–8
Solidago	2 feet	3–8
Sorrel	18 inches	3–7
Strawberries	18 inches	3–10
Sunchokes	2 feet	3–8
Tulips	12 inches	3–8
Thyme	9–12 inches	5–9
Walking onions	4 inches	3–9
Yarrow	18 inches	3–9

APPENDIX C: WHEN TO SOW FOR FALL AND WINTER HARVEST

Fall harvests are sown in mid to late summer as indicated below. For winter and spring harvests, subtract the indicated number of weeks from your Persephone start date to learn when seeds are ideally sown. For example, my Persephone period starts on November 30 (find your dates by searching for a day length calculator; there are several online. Persephone starts when day length drops below 10 hours). My winter-harvest spinach will be planted seven weeks earlier on October 12, while my spring-harvest spinach will be planted four weeks prior on November 9.

CROP	WHEN TO SOW SEEDS FOR A FALL HARVEST	WHEN TO SOW SEEDS FOR A WINTER HARVEST (WEEKS BEFORE PERSEPHONE)	WHEN TO SOW SEEDS FOR AN EARLY-SPRING HARVEST (WEEKS BEFORE PERSEPHONE)
Arugula	Late summer	8 weeks	5 weeks
Beets	Late summer	13 weeks	Not recommended
Broccoli	Late summer	Not recommended	10 weeks for sprouting varieties
Cabbage	Midsummer	15 weeks	Not recommended
Carrots	Midsummer	13 weeks	Not recommended
Celery	Midsummer	15 weeks	Not recommended
Chicory, radicchio, escarole, endive, and frisée	Late summer	10 weeks	7 weeks
Cilantro	Late summer	10 weeks	7 weeks
Collards	Midsummer	13–15 weeks	8 weeks
Cress	Late summer	5–6 weeks	5–6 weeks
Chard	Midsummer	10 weeks	7 weeks
Kale	Midsummer	15 weeks	8 weeks
Leeks	Early summer	20 weeks	15 weeks
Lettuce	Late summer	8–10 weeks	7 weeks
Parsley	Midsummer	12 weeks	10 weeks
Rutabagas and turnips	Midsummer	10 weeks	7 weeks
Onions, green	Midsummer	15 weeks	12 weeks
Pak choi and tatsoi	Midsummer	10 weeks	7 weeks
Spinach	Late summer	7 weeks	4 weeks
Mizuna	Late summer	8 weeks	5 weeks

ACKNOWLEDGMENTS

Thanks to Erin for showing me how to make gardens beautiful, and to Howie and Melany for their contagious excitement around ecological processes that seem insignificant (dare I say "boring") to the layperson. Beauty and science make me the gardener I am today.

Thanks to Carolyn and Kathy for providing opportunities to find my love for teaching. I would never have ventured to become an author had I not found so much joy in educating others.

Thanks to my kids, Owen and June, for their daily reminders of how little I actually know. Humility is a repeated theme in gardening and I'm grateful to have so much practice.

Thanks to Makenna, whom I'll forever credit with seeing something in me that I didn't see in myself—I've now written not one, but TWO books, Makenna!

And last, thanks to Lisa and Carleen, who waved their editorial wands over my discombobulated mess of a manuscript and transformed it into something worth buying and even sharing with friends. Editorial work is rarely appreciated for what it is: magic.

SUGGESTED READING

Bloom, Jessi, and Dave Boehnlein. *Practical Permaculture for Home Landscapes, Your Community, and the Whole Earth*. Portland, OR, Timber Press, 2015.

Coleman, Eliot. *The Winter Harvest Handbook: Year-Round Vegetable Production Using Deep-Organic Techniques and Unheated Greenhouses*. White River Junction, VT, Chelsea Green Publishing, 2009.

Coleman, Eliot. *The New Organic Grower: A Master's Manual of Tools and Techniques for the Home and Market Gardener*. 3rd ed. White River Junction, VT, Chelsea Green Publishing, 2018.

Cowden, Meg McAndrews. *Plant Grow Harvest Repeat: Grow a Bounty of Vegetables, Fruits & Flowers by Mastering the Art of Succession Planting*. Portland, OR, Timber Press, 2022.

Holderread, Dave. *Storey's Guide to Raising Ducks*. 2nd ed. North Adams, MA, Storey Publishing, 2011.

Jabbour, Niki. *The Year-Round Vegetable Gardener: How to Grow Your Own Food 365 Days a Year, No Matter Where You Live*. North Adams, MA, Storey Publishing, 2011.

Miessler, Diane. *Grow Your Soil! Harness the Power of the Soil Food Web to Create Your Best Garden Ever*. North Adams, MA, Storey Publishing, 2020.

Reich, Lee. *The Pruning Book*. Newtown, CT, The Taunton Press, 1997.

Ussery, Harvey. *The Small-Scale Poultry Flock: An All-Natural Approach to Raising Chickens and Other Fowl for Home and Market Growers*. White River Junction, VT, Chelsea Green Publishing, 2011.

Ziegler, Lisa Mason. *Vegetables Love Flowers: Companion Planting for Beauty and Bounty*. Minneapolis, Quarto Publishing Group, 2018.

METRIC CONVERSION TABLE

LENGTH

TO CONVERT	TO	MULTIPLY
inches	centimeters	inches by 2.54
inches	meters	inches by 0.0254
feet	meters	feet by 0.3048
feet	kilometers	feet by 0.0003048
yards	meters	yards by 0.9144
yards	kilometers	yards by 0.0009144

SOME COMMON CONVERSIONS

US (INCHES)	METRIC (CENTIMETERS)
1	2.54
2	5.08
3	7.62
4	10.16
5	12.70
6	15.24
12 (1 foot)	30.48

WEIGHT AND VOLUME

TO CONVERT	TO	MULTIPLY
pounds	grams	pounds by 453.5
pounds	kilograms	pounds by 0.45
gallons	liters	gallons by 3.785

INDEX

Page numbers in *italics* indicate illustrations and photos.
Page numbers in **bold** font indicate tables.

ENJOY YOUR BEST HARVEST EVER
with These Books from Storey

THE BACKYARD HOMESTEAD
edited by Carleen Madigan

From just a quarter of an acre, you can harvest 1,400 eggs, 50 pounds of wheat, 60 pounds of fruit, 2,000 pounds of vegetables, 280 pounds of pork, and 75 pounds of nuts. That's enough fresh, organic food to keep a family of four—year-round! This comprehensive guide shows you how to do it.

THE BACKYARD HOMESTEAD SEASONAL PLANNER by Ann Larkin Hansen

This practical workbook provides a detailed calendar of what to do on your small farm or homestead. For each critical turn of the year, Hansen offers an at-a-glance to-do list, setting your priorities for work in the barn, garden, orchard, field, pasture, and woodlot. Easy-reference charts, checklists, and record-keeping sections help you keep track of it all.

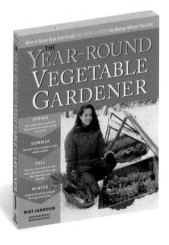

THE YEAR-ROUND VEGETABLE GARDENER by Niki Jabbour

Even in the heart of winter, you can harvest fresh produce! Jabbour shares her simple techniques for gardening throughout the year, covering how to select the best varieties for each season, the art of succession planting, how to build inexpensive structures to protect crops from the elements, and much more.